P9-CMR-439

TECHNOLOGY AND HUMANISM

WILLIAM G. CARLETON

TECHNOLOGY AND HUMANISM

SOME EXPLORATORY ESSAYS
FOR OUR TIME ~ *With a Foreword by*
MANNING J. DAUER

Vanderbilt University Press: NASHVILLE 1970

Copyright © 1970 by Vanderbilt University Press

The author and publisher express grateful acknowledgment to the editors of the following periodicals for permission to reprint essays which originally appeared in their publications:

"Ideology or Balance of Power?" *The Yale Review*, XXXVI (Summer 1947), copyright Yale University Press.

"The New Nationalism," *The Virginia Quarterly Review*, XXVI (Summer 1950).

"What Our World Federalists Neglect," *The Antioch Review*, VIII (Spring 1948).

"An Atlantic Curtain," *The American Scholar*, XXII (Summer 1953), copyright © 1969 William G. Carleton.

"Leninism and the Legacy of Western Imperialism," *The Yale Review*, LI (Summer 1962), copyright Yale University Press.

"Our Post-Crisis World," *The American Scholar*, XXXIII (Winter 1963–64), copyright © 1969 William G. Carleton.

"Centralization and the Open Society," reprinted with permission from the *Political Science Quarterly*, LXXVI (June 1960), pp. 244–259.

"The Growth of the Presidency," *Current History*, XXXIX (October 1960).

"The Ungenerous Approach to Woodrow Wilson," *The Virginia Quarterly Review*, XLIV (Spring 1968).

"The Revolution in the Presidential Nominating Convention," reprinted with permission from the *Political Science Quarterly*, LXXII (June 1957), pp. 224–240.

"Kennedy in History: An Early Appraisal," *The Antioch Review*, XXIV (Fall 1964).

"The Conservative Myth," *The Antioch Review*, XXVII (Fall 1967).

"Hawthorne Discovers the English," *The Yale Review*, LIII (Spring 1964), copyright Yale University.

"The Passing of Bughouse Square," *The Antioch Review*, XX (Fall 1960).

"Letter to a New PhD," *Teachers College Record*, LXIII (December 1961).

"American Intellectuals and American Democracy," *The Antioch Review*, XIX (Summer 1959).

"The Century of Technocracy," *The Antioch Review*, XXV (Winter 1965–66).

"Surrealism, Revolution, Reform—or What?" *The Virginia Quarterly Review*, XLV (Autumn 1969).

International Standard Book Number 0–8265–1154–6
Library of Congress Catalogue Card Number 70–112601
Printed in the United States of America by
Heritage Printers, Inc., Charlotte, North Carolina
Bound by Nicholstone Book Bindery, Nashville, Tennessee

843
.C28

For Vann, Glenn, and Peter

WITHDRAWN

124988

EMORY AND HENRY LIBRARY

❧ Foreword

■ THESE ESSAYS by William G. Carleton are of great importance because they analyze some of the most important trends for change in the twentieth century. The product of a truly seminal mind, they cover a surprising breadth of topics: some discuss the impact of United States technological development on social development, on government structure, and on our foreign policy. Other essays treat the development of the Cold War, still others treat intellectual developments at home and abroad, trends towards the development of pluralism in Communist countries, and the primary significance of nationalism in the developing countries.

A distinctive feature of these essays is the fact that, in a period when most scholars are becoming more and more specialized, Bill Carleton has been able to focus knowledge from many areas into an interpretative synthesis. This approach has been comparatively rare during the twenty-five years since World War II. Moreover, because of the "knowledge explosion" of the past two generations, to attempt such a synthesis is to run the risk of shallowness. Hence most interpreters have avoided broad-scale interpretation. But this overspecialization, among other problems, is now causing a justified revolt by a new generation. One reason for this revolt is that, especially in the United States, we have so dedicated ourselves to a cult of "presentism" that we now have lost a sense of direction. So, as the 1970s open, there is a general demand for more evaluation and more treatment of values. For these reasons the present collection of essays has a freshness which springs not only from the fact that important trends were discerned some twenty years before other observers discovered them, but also because they center on trends which are still the dominant and continuing ones of the seventies and after.

Why then are these essays so perceptive and still so fresh? The answer lies not only in the quality of mind possessed by Carleton, but also in his breadth of knowledge, in the extent of his reading, and in his judgment of cultural forces and of forces which move people. In history and the social sciences, Bill Carleton is broadly read. His love of history goes back to his childhood; he has read the chief historians and biographers from the classical period to the present. He also has read widely in the social sciences and in philosophy. And he truly is a humanist, a reader of the leading essayists and novelists ranging from the Americans to the British, to the Western continental writers, and to the Russians. Next he is an inquisitive traveler and observer, both in America and in Europe. All this has produced a uniquely humanist outlook which enables him to separate the wheat from the chaff, to determine what will be the consequences of developing human aspirations in a period of change and flux in technology, in economics, and in social structure. Finally he is gifted with a style which enables him to illumine topics clearly for the reader, topics which others have discussed in much more abstruse style. The outcome is a book which is informative, provocative, and a pleasure to read. There are points of interpretation or of emphasis which might be changed in the light of later events. But these are few. Much more often the bull's-eye is hit dead center.

The qualities of analysis possessed by Bill Carleton are familiar to those who have read his *Revolution in American Foreign Policy*, now in the fifth edition, published by Random House. This collection of essays is selected from more than one hundred articles which Carleton has published in periodicals over the past twenty-five years. A number of them have also appeared in various anthologies on foreign policy: for instance, the essay "Ideology or Balance of Power" has been reprinted in one of Norman J. Padelford's works on foreign policy and in eight other anthologies. Other essays have been reprinted in Asian or European periodicals or in anthologies on political thought, comparative culture, American politics, or creative writing. Thus the range of publication is broad.

Serving as a professor at the University of Florida for nearly thirty years, Carleton has provocatively stimulated college students with many of the ideas presented herein. He has also presented many of these ideas in lectures which he continues to give throughout the United States. The essays in this collection are representative, and through them there are certain unifying threads of thought.

Certain trends of change and development are discussed in these essays. In the first place, even before the Cold War began, Carleton was skeptical of the extent to which the diplomatic conflict between the USSR and the United States was predominantly an ideological one between Communism and capitalism. He recognized the primacy of the two super-powers but felt their opposition to be more that of the conflict over the balance of power. Furthermore, even in the 1940s he questioned the possibility of the threat of a disciplined monolith of Communist countries, and foresaw pluralism among the Communist states. Among the developing countries he felt that neither Communism nor capitalism would be controlling; rather he argued that nationalism would be the most important force among the newer national states. He also felt that in the long run there would be a third force composed of western continental Europe and England. He believed that this could be based on economic and technological forces rather than on legal structure, at least initially. Next he felt that while the wealth of the United States would have great impact, American leadership would ultimately be limited. To understand why, we may summarize the trends he sees in the United States.

Essentially the United States has power, springing from American technology. At home this industrial and commercial technology means a tendency toward increasing power by the national government. This national centralization is necessary because the American states have become units which are too small to be effective for many governmental problems. At the same time Carleton saw the impact of technology on the federal government as continuing to strengthen the role of the President as compared to Congress. Yet the very industrial and technological success of the United States has created a peculiar type of culture which raises difficulties for effective understanding of developing countries or of Europe. Neither the leaders nor the American public has the humanist background to comprehend trends in world development. This, Carleton believes, is a serious limitation on effective conduct of foreign policy. Also it creates dislocation in the United States and makes it harder for the American public to have a perspective on the problems of rapid change. The consequences are problems from a rootless culture. Clearly these are many of the central trends of the present. But to attempt further to summarize is to ignore shadings of meaning and nuances of analysis. Hence the reader is best advised at this point to turn to the essays themselves. There he will be richly rewarded.

University of Florida
January 1970

Manning J. Dauer

Contents

❧ A Word from the Author

■ PERHAPS THE distinguishing characteristic of these essays is their comprehensiveness. The time covered extends from the middle 1940s to the beginning of the 1970s. The matters discussed range from world crisis to bughouse squares, from Woodrow Wilson to the hippies of Haight-Ashbury. The subject of each article is broadly treated. The materials used are "interdisciplinary," and while the major approach is historical, additional sources are drawn on from political science, sociology, anthropology, psychology, literature, and other disciplines in the humanities. The point of view is catholic. Although my own values are not disguised, I have never been able to think of any one nation, race, religion, class, party, or ideology as embodying all virtue and wisdom, and hence I believe these essays have a rather high degree of impartiality, especially for this age of doctrinaire conflict.

What I have attempted to do is to cut through the tortuous complexities of our astonishingly fast-moving world to achieve perspective and to foreshadow the historical trends of today and tomorrow. I have done this in a rather free-wheeling fashion, probing the various possible alternatives—thinking out loud as it were—drawing liberally on the social and historical imagination. The work of the journalist is too restricted and immediate for this purpose, and very often the academician is too immersed in his specialty or too jealous of his professional reputation to allow himself such flexibility. One who does this inevitably runs the risk of making a fool of himself, and doubtless some of my judgments will turn out to be wide of the mark. On the other hand, reading over these articles after the lapse of time—in some cases several decades—I have persuaded myself that they are not without some insight and that if I have not always been right in discerning the prevailing tendencies and trends, I have at least raised a number of the decisive questions.

At the time these essays were written, I was not contemplating putting them together in a book; hence they exhibit some repetition, a number of inconsistencies and apparent inconsistencies, and a lack of systematic continuity. A bit of astute cutting and editing would have given them more unity and system—and "prophetic wisdom"—and the reader will note an occasional paragraph that I have added to comment on contemporary events. In general, however, I have preferred that they be published exactly as they originally appeared, warts and all, rather than that they be edited in the light of subsequent events. The freshly written introductions to each article, and some new footnotes, will point out some of the contradictions, iron out a number of them, bring matters up to date, and provide a measure of thematic continuity. The reader, though, is free to read the pieces in any order and as individual essays.

William G. Carleton

Gainesville, Florida
March 1970

I

World Politics

✑ Ideology or Balance of Power?

This article appeared in 1947, the year the Cold War came out into the open—with Russia and the Western powers going their separate ways over Germany and partitioning that country into a Communist part and a non-Communist part, with President Truman proclaiming the doctrine of containing Communism at the northern borders of Greece and Turkey, and with world politics becoming bipolarized on an ideological basis and the various countries aligning themselves into opposing blocs, one led by non-Communist America and the other by Communist Russia.

Had not ideology in the past played a subordinate role in the relations of countries to one another? Had not even the ideological Christian alliances formed to fight the various Crusades against the Moslems (1096–1291) foundered on the nonideological differences among the Christian allies? Had not Philip II's attempts in the late sixteenth century to weld Catholic countries against Protestant countries largely failed? Had not the post-Napoleonic ambitions of the autocrats of Austria, Russia, and Prussia to unite the absolute monarchies into a permanent Holy Alliance, as a basis for reactionary ideological co-operation in world politics, also failed; and had not nationalism grown apace?

Were we, then, in 1947, witnessing a dramatic revolutionary shift in the nature of world politics, entering a time when ideology would supersede nationalism as the primary force in international relations?

This article explores historically the nature of ideological conflict and how, in the international relations of the past, ideology has entwined with the national interest but has usually been subordinate to it. The article speculates on whether ideology is indeed becoming the primary element in world politics and, influenced by the climate of opinion of the time, tends to give an affirmative answer. The article which follows this one re-examines this question.

■ WHAT IS THE chief element in formulating the foreign policy of a nation? Is it ideology or is it balance of power? Is it both? If both, which weighs the more?

Did Sparta and Athens fight the Peloponnesian War because of the

The Yale Review, XXXVI (Summer 1947).

conflict of their social and political systems or because each feared the collective power of the other? Was Whig Britain at war with Bourbon France because of the rivalry of different institutions or because of the clash of competitive imperialisms? Did Burke's Britain fight Revolutionary France because of Tory fear of the Jacobins or because Britain feared for the balance of power? Did the United States go to war with Germany in 1917 and again in 1941 because of a conflict of cultures or a conflict of power?

What is this thing called "ideology"? There are, of course, ideologies and ideologies, and the term needs clarification.

Nations differ from each other in cultural and institutional patterns. But institutions inside nations are forever undergoing changes. These changes bring on institutional conflict within nations. Some fear these changes, while others favor them. Very often the institutional conflict taking place inside one nation is at the same time also taking place inside other nations. In other words, the struggle for institutional change often cuts across national boundaries.

In one age, the emphasis is on one aspect of cultural or institutional change; in another age, the emphasis is on another aspect. In the sixteenth century and the early seventeenth century, the chief institutional conflict inside countries involved religion. In some countries the Protestants won, and in other countries the Catholics won. But inside Protestant countries there were Catholic minorities, and inside Catholic countries there were Protestant minorities. Later, the chief institutional conflict was over constitutional and representative government. In some countries constitutional government won, and in other countries absolute monarchy held its own; but in countries where constitutional government won, there were still those who clung to the divinity of kings, and in countries where absolutism won, there were minority groups which favored constitutional and representative government. Still later, the chief institutional conflict was over political democracy. However, in countries where democracy triumphed, there were minority groups which opposed it, and in countries where democracy made little headway, there were minority groups which championed it. Today, in this middle of the twentieth century, the chief institutional conflict centers around the ownership of productive property. The struggle is one of capitalism versus some form of collectivism. Inside almost every country this struggle is taking place in one way or another. In some countries collectivism is further advanced than in others. Those in one country who favor collectivism find themselves in sympathy with those in other countries who favor it. Those in one country who oppose collectivism find themselves in sympathy with those in other countries who oppose it.

The decisive institutional conflict of any age has come to be known as the "ideological" conflict. While, in general, that conflict was in the seventeenth century waged over religion, in the eighteenth century over constitutional and representative government, and in the nineteenth century over political democracy, in the present century it is being waged over socialism.

That the institutional or ideological conflict is waged inside many nations at the same time, that the new institution struggling to be born is further advanced in some countries than in others, that this conflict cuts across national lines and leaves people and groups within the various countries with ideological sympathies for similar groups in other countries—all of this enormously complicates international relations. International wars often take on the appearance of deep-seated ideological conflict involving not only national interests but future institutional and cultural development. Men who belong to minority groups are often torn between two loyalties—loyalty to ideology and loyalty to nation. Such men may be damned as "Trojan horses" and "fifth columnists" if they follow their ideological impulse; they may be damned as disloyal to their ideology if they follow their national impulse.

What if a situation arises in international politics in which ideology and national interest do not coincide? In that case, which is put first by those who conduct the nation's foreign policy—ideology or national interest?

As we look back on the past, it seems that, when the hour of decision struck, the national urge was generally stronger than the ideological. During the Thirty Years' War, in an age when the chief ideological conflict was religious, Catholic France went to the aid of Sweden and the German Protestants in their conflict with Catholic Austria. To Cardinal Richelieu and other French leaders of the day it was more important for France to check Austria, for the Bourbons to humble the Hapsburgs, than for Catholics to stand together against Protestants. In the late seventeenth century and in the eighteenth century, it usually seemed more important to the absolute monarchs of Vienna and Berlin to check the French nation than that Hapsburgs, Hohenzollerns, and Bourbons should stand together in ideological alliance to maintain absolutism against the slowly rising tide of constitutionalism. Britain fought Bourbon France, Revolutionary

5

France, and Napoleonic France, for while the ideologies shifted and changed, the national rivalry and imperialist conflict did not. In more recent times, Tsarist Russia has been allied with republican France, democratic Britain with oligarchic Japan, democratic America and democratic Britain with Communist Russia, and Communist Russia with Nazi Germany. . . . National interests make strange bedfellows—in the face of a threat to the balance of power, nations will make alliances with the ideological devil.

Where, in general, the dominant ideology within the nation and the national interest do coincide, in what direction have the ideological minorities usually gone? In a crisis, what have these minorities done—have they followed their ideologies or their national patriotisms? In the past, the tug of national loyalty has generally won out over ideological persuasion. British Catholics rallied to Elizabeth in her fight with Catholic Spain. Charles James Fox's Whigs rallied to Pitt in the national fight against Napoleon. And in the two global wars of this century the Socialists did what they boasted they would never do—in the main, they rallied to the national standards. A Coriolanus and an Alcibiades, a Vallandigham and a Petigru, a Quisling and an Einstein, a Laval and a Thomas Mann are relatively rare in history. (This is, I know, an incongruous list of saints and sinners, but they have this one thing in common: in one way or another, all sacrificed homeland for ideology.) And should all our hopes of peace collapse and the future see a third world war involving the Soviet Union against the United States, it is still a safe prediction that Eleanor Roosevelt and Colonel Robert McCormick, Max Lerner and Hamilton Fish, Philip Murray and John Bricker, Henry Wallace and Lawrence Dennis would be fighting side by side in the name of national interest and national survival.

However, it would be a mistake to come to the conclusion that because ideology has not played the leading part in historic international relations it has played no part at all. It has played its part, an important part. Where national interests and dominant ideology within the nation coincide, a national war can be made to appear an ideological one, morale can be strengthened, and enthusiasm intensified. If a nation possesses considerable ideological unity, it will be in a stronger position to win a war; if a nation's enemies are ideologically divided, those enemies will be more susceptible to fifth-column tactics. When a country is rising to challenge the old balance of power, nations thus threatened will make alliances more easily and earlier if they have similar institutions and cultures; alliances will be more difficult and will come later (perhaps too late!) if the nations thus threatened have dissimilar institutions and cultures.

Just as national interests and balance-of-power considerations seem to have been the most important causes of war, so also they seem to have been most affected by war. The results of international wars seem to have been more significant in their national and balance-of-power aspect than in their ideological aspect. The Grand Alliance against Napoleonic France checked France and saved the European balance of power but did not succeed in arresting the spread of revolutionary ideas. The Grand Alliance won the war nationally but in the end lost it ideologically. However, when nations outside of France became more and more influenced by French Revolutionary ideas, those ideas took on the appearance of being their own—they were assimilated into their own national cultures. Again, at the close of the First World War, it appeared that middle-class democracy would triumph in the world, but now we can see that instead, the First World War marked virtually the end of the advance of middle-class democracy in any large areas of the world. However, the First World War did parry the German threat to the balance of power; it did temporarily satisfy the national interests of the victors; the war was won nationally even though it was lost ideologically. Again, the victory of the United Nations in the Second World War saved the world from a second German threat to the balance of power—that much is certain. Perhaps it will also have hastened socialism, but, in any event, the ideological results are not so clear as the national ones.

It seems, then, that the influence of great wars on the cultural and institutional trends of the time is exaggerated. These trends arise out of conditions and forces operating within the nations. Wars affect them. Wars may accelerate these trends or slow them up. Wars hardly create them.

The Marxists would claim that what appears to be the national interest has in fact been primarily the interest of the dominant class in control of the state and that class interest has governed the foreign policy of national states and involved national states in wars for class ends. The great national wars, according to Marxist doctrine, have been wars in the interests of the dominant economic classes in the warring nations—conflicts of rival imperialisms. Where, in the past, ideology has seemed to split the dominant class, such ideological conflict, according to the Marxists, has been superficial or at least secondary to that class's economic interest. When a threat has appeared to the economic interest of the dominant class, these secondary ideological ranks have been closed in the class interest masquerading in the name of the nation. Of course, say the Marxists, the Catholic landlords and businessmen of Elizabeth's day rallied to the fight against Catholic Spain because their profits were involved. Of course, the Marxists say again, the Charles James Fox Whigs, enamored though they were of many of the principles of the French Revolution, rallied to the war against Revolutionary France because their dividends from the British East India Company and from many another chartered company were at stake. (They fail to recall that the peace party in Britain led by Fox did not actually "rally" to the war until France under Napoleon had threatened to control the Channel and invade Britain.)

To be sure, the great conflict *is* ideological, say the Marxists—an ideological conflict between the exploiting capitalists and the exploited workers—and when this ideological conflict comes to the fore and cuts across national states, then the conflict between nations will be seen to be a sham and a swindle, the ideological factor will become stronger than the national factor in international relations, national wars will be converted into civil wars, and the erstwhile dominant economic class, now embattled, will find it more and more difficult to cloak its class interests in the garment of nationalism.

It may be that the Marxists are correct as to the future. It may be that the ideological will supersede the national as the number one factor in international relations. It may be that now the great institutional conflict within nations has come to be directly economic (and not indirectly so, as in the case of many of the ideological conflicts of the past) and the fundamental issue easier for all to see, that more and more will men respond to the class and ideological appeal rather than the national appeal. And if men more and more are learning to think as the Marxists do (the postwar elections in Britain and especially in Continental Europe indicate that they are) then it is quite possible that our century will see international relations conducted in fact and in name more along ideological than national lines. Indeed, the Second World War was in many ways a series of civil wars, with Socialists and Communists in Axis countries praying for an Allied victory and fascists in Allied countries supporting collaboration with the Axis. Ideological minorities were co-operating with the national enemy on a scale never equaled in the days when Protestantism or constitutional government or political democracy was the pivot of ideological conflict—though perhaps some of this co-operation should be credited to the increase and improvement in the means of communication. And today, Henry Wallace speaking to Britain in a way hostile to the views of a majority of his own countrymen or Winston Churchill speaking to America in a way hostile to the views of millions of left-wing Britons is another example of how in our time ideology cuts across national boundaries.

At present, the United States and the Soviet Union face each other as the predominant powers of the earth. Each is viewing world politics in a different light from the other.

Our policy-makers are thinking more in the old terms of nationalism, self-determination, and the balance of power. Soviet influence on Communist parties the world over is looked upon largely as old-fashioned aggression and imperialism. There is widespread fear that Russia as Russia will upset the balance of power in Europe and Asia. Poland and Rumania and Bulgaria and Yugoslavia and China are appealed to by our government on the basis of national self-determination. (National self-determination, once a liberal rallying cry, has become also a conservative shibboleth.)

The Soviet leaders, on the other hand, are thinking more and more in terms of ideological conflict, class warfare, social politics. To Soviet policy-makers the contest in the world today is not so much one between the United States and Russia as it is between world fascism and world Communism. And those in other countries who follow the Soviet lead take much the same view. Communists and extreme leftists the world over are giving their allegiance not so much to nations as to ideology. Russia is merely the instrument to be used in spreading ideology. Moscow influences Communist parties in all countries, and Communist parties everywhere influence Moscow. There is common indoctrination, consultation, co-operation. It is a two-way street, though the pre-eminence of Russia in the movement makes the outgoing counsel from Moscow weightier than the incoming counsel.

If a third world war should come—a war between the United States and the Soviet Union—for millions of Communists the world over that war would be viewed almost exclusively as an ideological one. In countries where the extreme left is in control, national policy would be made less on the balance-of-power idea and more on ideological considerations. In the United States, where Marxist ideas have scarcely penetrated at all, the war would be viewed in the old nationalistic terms. It would be a war to protect our shores from invasion, to uphold the balance of power, to save the world from Russian domination. As has already been said, in a

9

final showdown nearly all of us would stand together. Our Communist fifth column would be smothered by the avalanche of nationalistic sentiment. American psychological warfare abroad would naturally reflect our own view of the situation, and we should probably err on the side of making too much appeal to nationalistic sentiments, which in large areas of the world would be effective only in rightist circles. Our own failure to take into account the enormous strides the ideological point of view has made among the masses of people in Europe and Asia in recent decades, and particularly since the Second World War, might betray us into making very serious mistakes in the conduct of that war.

Should the Communists eventually win control in important countries outside Russia—in China, in Germany, in France, in Italy—it is quite possible that ideology would triumph over nationalism and a new international state emerge. After all, the nations themselves were built not so much on contract as on class and functional foundations. The feudal states of Western Europe were put together into national states by the rising bourgeoisie joining hands across the boundaries of feudal provinces and communes. Just as the national state (which sometimes cut across nationality) was made by the bourgeoisie, so a new international state may be made by the co-operation of leftist parties joining hands across national boundaries. If an international state should come about in this way, then, of course, the old balance of power as played by national states will be relegated to the historical limbo. The Communists still confidently expect that Communism will lead to such an international state.

Of course, the Communists may be wrong about this. Nationalistic forces may be stronger than Communist intentions. When the Communists come into actual power in a country they inherit the national paraphernalia—the nation's history, culture, aspirations—and concessions have to be made to them. The Russian Communists have made many concessions to nationalism. The Chinese Communists have often pursued a course independent of Moscow, and if they should gain actual power in all China they might have to follow an even more independent course. Thorez, Communist leader in France, says over and again: "Different countries, different methods." And the moderate Socialists, even more than the extreme Socialists or Communists, have, of course, made many more concessions to nationalism and have proclaimed this as a virtue. Moreover, if the United States should use the moderate left (Socialists) to check the extreme left (Communists), such a policy might help to prevent a third world war and also help to halt a trend towards a Communist world state with the Soviet Union as its nucleus. Then, too, it is even possible, as Edward Hallett Carr has suggested, that in a state where work and wages and social security are dependent upon government, where the mass of people will seem to have as direct a stake in the national government as at one time only the bourgeoisie enjoyed, the mass of people may feel even more keenly their tie to the national government, and socialist states might end by being as nationalistic or even more nationalistic than bourgeois states. If this should turn out to be the case (although the Marxist analysis that private ownership of the productive processes and private profits are the basis of imperialism and international conflict may prove to be sounder) then socialism, like the commercial revolution and the Protestant Reformation and the French Revolution and the early stages of the industrial revolution before it, will have ended by intensifying nationalism, in spite of the aims and intentions of many of its disciples. And then, should socialism spread and thus turn out nationalistic, the old historic pattern of the national balance of power would continue the dominant role in international relations, even though Socialist and Communist states remained nominal members of an international organization like the United Nations.[1]

Of one thing, I think, we can be sure: even the creation of an international Communist state would not end the conflict over power. Within the international state would be different social and economic groups with different interests. They would fight for control, just as groups and classes within national states now fight for control. The international state would end the national balance-of-power conflict, but it would not end the ideological conflict inside it. The ideological conflict of the future will be different, but it will continue in some form. Even if the Marxists should achieve one half of their ideal, the international state, they would

1. Throughout this book, Socialism and Socialist have been capitalized when they refer to definite, articulated, and organized political movements looking to a nontotalitarian collectivism; but they have not been capitalized when the reference is to more formless collectivist movements, or when it is used in a generic sense to include all kinds of collectivisms and collectivists ranging from unorganized to highly organized, from moderate to extreme, from nontotalitarian to totalitarian. This distinction is often very fine indeed. On the other hand, Communism and Communist have been capitalized uniformly (except in those rare instances when the reference is to a vague, general, and even utopian communism), for the Communism and the Communists important in the world politics of today are always highly organized and self-conscious, whether the reference is to Soviet Communism or to Maoism or to Titoism and its variations.

not succeed in achieving the other half—the abolition of all its internal group differences. Within the international state would be geographical sections, cultural diversities, a wide gamut of different industries and economic enterprises, and various social classes: commissars, managers, engineers, technicians, skilled workers, unskilled workers, farmers, and so forth. These various factors and groups would generate conflict over government policies and there would arise struggles for power which even Communist purges could not keep down. These struggles would cut across the old national boundaries.

In short, evidence is already strong that the Marxists may be able to make the ideological conflict rather than the national conflict the pivot of mid–twentieth-century international politics. As a result of this ideological conflict the more extreme Marxists—the Communists—may even be able to establish an international Communist state, but this is far less certain either because they may not win the ultimate international conflict even after succeeding in making the issue an ideological issue, or because the forces of nationalism may be too strong for the Communists once they inherit the various national governments. But even should the Communists win and establish an international Communist state, it is almost certain that new social and economic groups would arise to continue the struggle for power within the international state, peacefully perhaps, but a struggle none the less. If so, then national conflict over the balance of power would disappear, but ideological conflict within the international state would persist.

Should there be a third world war and should the United States, using the old shibboleths of nationalism, win the war, it is doubtful even then if political nationalism and the balance of power would continue as they have done in the past. We are perhaps too close to events of the last thirty years to be able to see just how far the old foundations of nationalism have already been eroded away. In the fifteenth century, participants in the wars of Louis XI and in the Wars of the Roses were too close to those events to realize that feudal power was being overthrown before their very eyes and a strong national power built. So probably with us today. The historians of the twenty-second century, more clear-eyed than we, may look back and see that the great international wars of the twentieth century were in fact dissolving nationalism and building internationalism.

For nationalism today is truly beset from all directions. Among the dissolving agents are: the cumulative impact of technology and science resulting in the continued drastic elimination of distance and space, the atomic bomb, the release of atomic energy and the overwhelming necessity of having to extend international functionalism to control what in the future will probably be the world's most important source of industrial production; the palpable absurdities of fascist nationalism and the revulsion from fascist nationalism even in countries which experienced it; the realistic education given millions of men who participated in the great wars of the twentieth century in areas remote from their homelands; the glaring fact that real national power in the international power conflict is today possessed by only two nations, leaving all other nations as mere outsiders with no important power; the growth of international cartels; the Socialist appeal to the brotherhood of man; the Communist insistence on international action and the actual co-operation of Communist parties from country to country; the socialist outlook of many of the nationalist leaders of the colonial countries in revolt against imperialism; the body of practical experience in international co-operation gained through the League of Nations and the United Nations. The Communists would put the world together through something like a Communist International; a majority of the people of Britain and the United States would prefer to see the world put together by the slow functional growth of an organization which originated in contract—the United Nations. (Inside the United Nations, too, there would, of course, be power politics, but as international functionalism grew it would come to be less and less power politics based on nationalism and more and more power politics based on ideology, that is, group and class conflicts cutting across the national lines.) And there are those, like James Burnham and the oversimplifiers and distorters of Arnold Toynbee, who are so impatient with the slower and wiser methods of bringing an international state that they would have America attempt to build a "universal empire" to end "a time of troubles." Whatever method of putting the world together will ultimately prevail is still anybody's guess, but that the world in our time is in painful process of being put together is more than a guess—it is a hypothesis based upon a growing accumulation of evidence.

Thus anyone called upon to answer the crucial question in international relations today would be, I think, on safe ground in saying that, from the rise of national states and up to about now, the chief element in international relations has been nationalism and the national balance of power. But he should warn the questioner not to be misled by this historic fact or by the superficial aspects of the present diplomatic duel between the United States and the Soviet Union, especially as that duel is generally interpreted in the United States. Because this middle of the twentieth century may be witnessing the epoch-making shift in the foundation of international politics from the nationalistic balance of power to ideology, evidence of which we shall ignore at our peril.

◆ℰ The New Nationalism

Only three years after the appearance of the preceding piece, the author published the following article, which clearly resolves the question of the relative importance in world politics of ideology and balance of power in favor of the latter—in favor of nationalism and the national interest. This view was anticipated by the author even earlier, when in the Summer 1949 number of the *Virginia Quarterly Review* he published an essay entitled "Is Communism Going National?" That article attempted to point out the enormous implication of Titoism, its portent of a growing nationalism within Communist countries.

It has taken the makers of American foreign policy and the American public a long time to recognize that a monolithic international Communism is largely a myth. Only today, twenty years later, are Americans beginning to show some awareness that Communist countries do not stand together on all, or even most, questions of international politics. In 1968, a United States Senator, the venerable George D. Aiken of Vermont, stated on the Senate floor that international Communist solidarity was "an invention," "a self-destructive fantasy." It took courage, even at that late date, for an American politician to make such a statement publicly.

■ IN SPITE OF strong indications of the persistence of nationalism, there is a general notion, part belief and part hope, that nationalism is gradually diminishing, that little by little it is being replaced by internationalism. This notion was particularly strong near the close of World War II; it was fed by the collapse of fascist nationalism, by the enthusiasm of victory won by a widely inclusive international coalition, and by the high hopes entertained for the germinating United Nations. During the past five years the world situation has become clearer, and we are in a better position to appraise future trends in international relations.

It now seems fairly certain that nationalism will again dominate the twentieth century as it did the nineteenth century. True, in Western Europe, where modern nationalism first developed and grew to maturity, it seems to be weakening. A number of factors are here at work to produce this result: a drastic decline in strength due to the frightful losses involved in bearing the brunt of two great nationalistic wars; the collapse of old-fashioned imperialism, the loss of overseas colonies, and the withering of foreign investments; the distressing realization that national power is passing from the nations of Western Europe to nations in other parts of the world. Still, even in Western Europe, in spite of notable advocacy and the immense practical advantages to be gained, no really effective steps have been taken toward the formation of a federal state. And in Eastern Europe, in Africa, in the Americas, and in Asia nationalism seems to be definitely on the rise. Indeed, the coming of the twentieth-century industrial revolutions to the backward areas of the world threatens to build and intensify nationalism in those areas in much the same way that nineteenth-century industrial revolutions built and intensified nationalism in the older industrial countries.

There is a facile and an uncritical assumption that as cultural barriers are broken down, as the various parts of our planet come to be more and more alike, as the industrial and technological and scientific revolutions increase commonalities, as more peoples come to be urban and metropolitan, the world will move toward political unification. This may be the ultimate result, but progress toward political unification may not be much greater in the twentieth century than it was in the nineteenth century. Indeed, since the breakup of the medieval order the world has been getting closer and closer physically and culturally. The commercial revolution, the rise of capitalism, imperialism, the scientific revolution, and the industrial revolution have all contributed to the eroding of cultural diversities. But, paradoxically enough, as cultural diversities diminished, political nationalism increased. The industrial revolutions, coming to different countries at different times, were developed within national patterns, and consequently national unity and the feeling of national self-interest were increased. And this process of breaking down cultural diversities while at the same time intensifying nationalism—this process characteristic of the sixteenth, seventeenth, eighteenth, and nineteenth centuries—now shows definite signs of continuing through the twentieth century.

Liberal thinkers of the nineteenth century thought that nationalism could be curbed by law and by an international organization based upon contract. These conceptions suffered a severe setback when the League of Nations foundered on nationalistic difficulties and fierce ideological conflict. They are now being tested in the United Nations, and again national difficulties and fierce ideological conflict threaten disaster. Another unfavorable portent is the failure of functional group alliances to cross national frontiers and to cooperate with similar groups abroad and against dissimilar groups at home—and thus apply the process of group conflict and compromise of domestic politics to international politics, disrupt national unities, and build a going international society. Delegations in the United Nations still operate as solid national fronts, and the United Nations is thus largely an arena for the playing of national balance-of-power politics. There are, of course, some signs that leftists are crossing national frontiers to co-operate with leftists, centrists with centrists, rightists with rightists; but this tendency is still too weak to counteract in any effective way the forces making for a continuation of national solidarity. Besides, this process, even if it developed appreciably, could operate for all groups only among democratic countries; dictatorships would not allow it, except for the particular ideological groups the dictatorships represented.

And this brings us to Marxism. The Marxists have claimed that when one group—the class-conscious proletariat organized as a Marxist party—came to power it would cross national boundaries and join hands with similar groups that had come to power in other countries, and thus the international proletariat would end the national state and create the international state. The Marxists contended that the national state was the creation of the bourgeoisie, that the national state was a broad cloak to disguise bourgeois interests, that through nationalistic imperialism and nationalistic wars the owners of production and their parasites exploited the masses and served their own class interests. Therefore, predicted the Marxists, when the workers came to power in old national states they would destroy those national states and establish an international workers' state.

This Marxist conception of nationalism has fooled and is fooling non-Marxists of the West today. Many non-Marxists of the West think that this Marxist concept is historically true; and more, that Communists today subscribe to it in the old literal, orthodox Marxist-Leninist sense and always operate on it in the realm of practical politics. Hence many in the West who have had a hand in American foreign policy have fondly believed that the non-Communists had a monopoly on nationalism; that the wars of national liberation in Asia would have to be non-Communist revolutions because Communism by its very nature could not be nationalist; that nationalist revolutions would be bourgeois simply because they were nationalist.

17

The truth is that contemporary Communism does not conform strictly to the Marxist-Leninist dogma with respect to nationalism and internationalism. Stalinism at home is more national than the Marx-Lenin formula contemplates, and it makes allowances for social revolutions waged as wars of national liberation. In actual practice Communist revolutions more and more take on the nature of nationalist revolutions, particularly in Asia. And as a matter of fact, the whole Marxian conception of nationalism as the mere creature of the bourgeoisie is much too simple and is not borne out by anthropology and history.

The essential characteristics of nationalism are older than modern nationalism, older than capitalism. To date nationalism from the commercial revolution, from the rising power of the modern bourgeoisie, is inaccurate. Men have always made a marked distinction between the in-group and out-groups, and loyalty to the in-group, whether it be tribe or city-state or nation, has been universal and persistent. (There has been, too, a continuing historical process of combining out-groups with in-groups to form larger in-groups, and it is this process that gives the adherents of internationalism their hope of ultimately breaking down nationalism.) Imperialism, too, is older than modern nationalism, older than capitalism. Indeed, imperialism was one of the most prominent aspects of antiquity. It is true, of course, that every great movement of modern times—the commercial revolution, the Protestant Reformation, the French Revolution, the rise of democracy, the industrial revolution—has intensified nationalism. And paradoxically enough, the modern Socialist and Communist movements, by bringing direct government services to the mass of peoples, may also result in strengthening nationalism, in spite of the hopes, the predictions, and the dialectical demonstrations of their prophets.

National differences represent more than the mere rivalries of conflicting businessmen. Different nations *do* have different needs and different cultures. They represent different geographies, different economic necessities, different stages of historical development. They are "culture-bound" in different ways. Even the coming of machine civilization does not break down all cultural differences, and it tends to lessen power differentials and thereby to intensify the power conflict.[1]

1. This latter phrase must now be modified in the light of the subsequent nuclear revolution in military technology, which has enormously widened the power differential between the supernuclear powers and the nonnuclear powers.

Both in Europe and in Asia, the patent truth is that today Communist revolutions are taking place within the framework of individual countries and are being established and consolidated along national lines. These revolutions are not so much international as they are national; the people look for work, and social services, and consumer goods, not to an international Communist state, which is mythical, but to individual national governments, which are real.

A genuine international revolution creating a genuine international state was expected at least four different times in this century: at the outbreak of World War I, at the time of the Bolshevik Revolution, at the end of World War I, and in the concluding days of World War II. Each time these expectations miscarried, and as a result Communist revolutions have been forced to develop within national patterns.

When the truly international revolution failed to materialize in 1919 and the early 1920s, following World War I, Russian Communism itself became more national and developed Stalinism as distinguished from Marx-Leninism. The sacred texts, of course, were not repudiated, but in practice they have been watered down considerably. In Russia a totalitarian Soviet patriotism has been evolving. For purposes of fomenting revolution abroad, the antithesis between bourgeois nationalism and proletarian internationalism is no longer so clearly drawn; national and social liberation are viewed as one and the same process to be achieved under the leadership of the Communist party. But even this watered-down version does not go far enough to meet nationalist realities, and increasingly Communism seems to divide between those who follow Moscow rather closely and those who would take a more independent national course, as for example Tito in Yugoslavia. Today the Communists, like the Socialists in the period around World War I, seem to be dividing between the internationalists (whose internationalism already has been watered down) and the nationalists.

It is probable that we have been thinking of Socialism and Communism too much as complete social systems and not enough as mere economic systems. Perhaps it would be nearer the truth to think of them as mere economic systems capable of being democratic or totalitarian, international or national, peaceful or imperialistic, depending upon the larger cultural milieus within which they work. Today they are working within national milieus and hence cannot escape national expression.

It is even possible that in Asia modern nationalism will emerge for the first time with the Communist revolutions just as in Europe modern nationalism emerged for the first time with the capitalist revolutions. Communism and nationalism are being linked together in wars of national liberation, just as democracy and nationalism were linked together in Europe's nineteenth-century wars of national liberation. For instance, the Chinese Communists are making the political unification of China one of their primary goals. Real nationalism in China may come to be as inseparably connected with the Communists as in its formative stages real nationalism in Europe was inseparably connected with the rise of the bourgeoisie. In Asia, Communism may be the maker of nationalism, not its destroyer. Under Communism, the mass of Asiatics in some countries may come into close contact with their governments for the first time in their history, and this contact, because of the many functions exercised by Communist governments, will be far more intimate than the contact of the mass of Europeans with their national governments in the days when European nationalism was emerging. And the industrial revolutions engineered today by the Communists are being worked out within national patterns, just as were the industrial revolutions led by the bourgeoisie in the nineteenth century.

What of the solidarity of the proletariat? Even as a revolutionary force it has been exaggerated, particularly in its international role. And after a successful Communist revolution within a nation, will it survive? Or will not the successful revolutionaries re-emerge as different and proliferating groups and classes? Will not the new Communist societies develop new groups and classes? Will not managers, directors, engineers, technicians, professional classes, skilled workers, unskilled workers, and farmers evolve within the Communist states, breaking the so-called solidarity of the international proletariat and making impossible any real international action by an international proletariat—in the inclusive sense of the old Marx-Lenin formula—across national boundaries? Will not the old international proletariat then become a revolutionary myth rather than a reality for practical international co-operation and action? In short, will not the so-called international proletariat inside Communist states turn out in fact to be neither international nor even proletarian?

It is probable that today the Communist world owes much of the co-operation that does exist among its various national parts to its own feeling of precariousness and to its fear of the non-Communist world. Should Communism grow and spread, should the opposition shrink and grow weaker, undoubtedly national rifts inside the Communist world would sharpen and multiply.

Nationalism is growing among the backward peoples everywhere, not only because it is an instrument to drive out ruling colonial powers, but also because only centralized state machinery can satisfy the mass demand for modern technology, heavy industries, and the rising standard of living these make possible. The backward peoples, led by their native intellectuals, want the industrial revolution, and they want it now—in this generation. They have no native capitalist class or middle class sufficiently large to finance modern technology, and they do not want the increased wealth produced by industrialization to be siphoned off as profits for foreign capitalists. They want industrialization; they want it now; and they want its benefits conserved for their own societies. Only centralized state power, only collectivist organization in some form, only government financing can bring in this generation so gigantic an economic transformation. Every vital revolutionary movement among backward peoples envisages industrialization and a great increase in centralized national power; every vital revolutionary movement is both social and national. This is true whether the leader is Sun Yat-sen or Mustafa Kemal or Nehru or Tito or Mao Tse-tung. The revolutionary movements in backward countries differ only in degree of socialization, centralization, and freedom; in the allocation they propose to make of the new national wealth; in their various attitudes toward Soviet Russia.

Nationalism, then, under different names, is on the march among the backward peoples. But it is not bourgeois nationalism. Indeed, its primary urge comes from the *absence* of a bourgeoisie. It is because the native middle classes are insufficient to finance industrial revolutions that all these movements are under compulsions to increase and socialize national power. (In this connection it is also significant that even in the West the strongest proponents of nationalism are no longer the bourgeoisie. For instance, in the United States during the nineteenth century it was the bourgeoisie who spearheaded centralized national power, whereas in the twentieth century it has been the labor groups who have spearheaded it.)

The very nature of the situation in backward countries gives the Communists certain advantages. The Communist fatherland, Russia, is the example inspiring all backward peoples. These people see in Russia a once-backward people who in one generation lifted themselves from medievalism to a degree of modernity. The Communists, too, are better organized, assert a ruthlessness over even workers and peasants which facilitates change and the accomplishment of specific goals, and are fired by the proletarian myth. But the Communists suffer some disadvantages, too. There is the fear of becoming an appendage of Russia. There is the conflict of native nationalism and the old myth of proletarian internationalism, still not modified enough to suit the views of many nationalists. Nehru, for instance, is reputed to hold the opinion that the Communists are still not nationalist enough to satisfy the growing nationalism of Asiatic peoples. The Communists are in something of a dilemma. If they do not espouse the cause of nationalism, they will not get into power in the backward countries; if they do espouse it, they endanger still further Communist internationalism (or Russian imperialism). But no matter what the original intentions of the Communists, when Communists acquire national power they do not seem able to escape going more and more national.

Americans must awaken to the fact that non-Communists have no monopoly on nationalism. The Communists, too, may use that weapon, and use it against us. And if in playing for the high stakes of international politics we sponsor nationalist revolutions in Asia and in other backward areas, we must reconcile ourselves to the fact that these revolutions will be in large measure socialist revolutions and not bourgeois or status quo revolutions. For today nationalism and socialism and nationalism and Communism are not incompatible; indeed, the great wars of national liberation in our times are not only nationalist but also socialist (as in India and Burma) and Communist (as in China and Indo-China).

If nationalism is rising over those extensive areas of the earth we have hitherto called "backward," then the implication of this for future international relations is most significant. The ultimate effect may well be to halt the Soviet-American polarization of power and to revive the old multiple balance-of-power system. Also, internationalism by way of long-time American or Soviet empire, and internationalism by way of Socialist or Communist international solidarity may have to be discarded as probabilities for the future. Our hopes for political internationalism may have to center around the United Nations, frail reed as this may appear to be in a world of old bourgeois nationalism and of spreading and intensifying collectivist nationalism.

✒ What Our World Federalists Neglect

There has always been a widespread notion in the United States that the creation of political entities and governments is a relatively simple matter. Did not the Americans, having decided that the time had come to have a national government, send to Philadelphia in 1787 a body of delegates who "struck off" a Constitution and created a nation? Could not an international government be created in some such similar fashion? Reflecting this naive belief, many Americans, following World War II, advocated "World Federalism" and "Union Now." The following article was written to examine critically the views of the exponents of World Federalism and of Union Now.

To be sure, America's federal government is founded on contract, but before that contract could be written, ratified, and implemented, and in order to sustain it over the decades, certain functional political forces had to operate. Contract and law alone, without a life-giving politics, would never have been enough, and the same is true if world government is ever to become a reality. It is strange that so many Americans have overemphasized their contractual experience and ignored their even more important political experience which has made their Constitution—that is their contract—a practical success.

The following article refers repeatedly to the co-operation of Communists across national boundaries. At the time the article was written, international Communist co-operation was much more of a reality than it is today, for since that time Titoism has spread, China has gone Communist and refused to follow Soviet leadership, and during the 1950s and 1960s nationalism within Communist countries asserted itself more and more.

The chief barrier to the germination of a viable international federalism still stands—the fact that all Communist governments are dictatorships, never allow internal dissident groups to become articulate, and would certainly never permit such groups to co-operate with like interest groups of other countries within an international organization.

■ ONE THING and only one thing will make possible the kind of international government for which Americans today are groping,

The Antioch Review, VIII (Spring 1948).

and that is the gradual displacing of national conflict by the group conflict that cuts across national boundaries.

At first blush this may sound shocking to most Americans because it
involves the substitution of one kind of conflict for another and because
it suggests the class and group struggle most Americans would like to
ignore. Yet this is not Marxism but is instead in the very best traditions
of American federalism. (James Madison, for instance, saw clearly the
group struggle long before Marx.) And while it perceives the continuation
of conflict—group and ideological conflict rather than national conflict—
it also perceives the decisive difference between them: national conflict frequently leads to full-dress wars while group conflict seldom does.

Paradoxically enough we can be saved only by conflict, for only as
group conflict expands and extends to international politics will national
conflict diminish. Genuine international government based on consent
will come only with the shattering of the unity of nations and the *reciprocal* willingness and freedom of social groups in one country to join
hands with similar groups in other countries, even in opposition to dissimilar groups in their own countries, to accomplish common group purposes felt to be superior to national purposes.

We have become accustomed to the weary stereotypes that beg the
real question. What are some of these stereotypes? That we should "think
internationally." That we should have the "will" to make an international
organization work. That we should "curb nationalism." That we should
"limit national sovereignty." But even an international body which had
the legal power to exercise a broad measure of the sovereignty now exercised by the national states would not result in real international government unless the national blocs gave way to international divisions
which crossed national lines. So long as each nation came to the international body seeing all questions from the national point of view and
discussing, arguing, voting on all questions as a national unit vis à vis
other national units, no genuine progress toward internationalism would
be made. We would merely have created another arena in which the
conflicting nationalisms could air their differences while at the same time
retaining the national solidarity which would enable them to settle those
differences by resort to war.

Of course the international body must have enough sovereignty assigned it to allow it to deal with some of the important matters that are
clearly seen to affect the everyday lives of people in their group and
class relationships. Otherwise people will never take enough interest to

divide along group lines that cross the old national boundaries. But the mere creation of an international organization based upon contract is not enough. The mere yielding of national sovereignty is not enough. Within such a legal structure must be the willingness—indeed, the urgency—of large numbers of people inside one country to co-operate politically with like-minded people inside other countries and to stand against people in their own country who differ with them on vital international questions. What is needed is not the mere curbing of political nationalism but its disintegration. An international organization will develop into real international government only to the degree it is implemented by functional conflicts that cross the old national barriers. The extent to which people do or do not see this, the extent to which they want or do not want to see this—indeed, the extent to which they are or are not already doing this—is the extent to which we are or are not ready to make a beginning—a mere beginning—toward international government.

II

What gave reality to the national state in the years of its development was not law or government structure so much as functional co-operation. True, the national states of Western Europe were put together around the legal theory of the sovereignty of the king. But for centuries this theory remained a mere fiction while real sovereignty was wielded by the feudal states and medieval communes which went to war with each other just as do the national states of today. But gradually the emerging bourgeoisie, because of class and group interests, joined hands across feudal states and communes to give real power to the king. As the sovereignty of the king became a reality, as the king became the real source of power, the old aristocracy also banded together to control the king and to bend government to its interest. Thus as class lines crossed feudal and communal boundaries the national state evolved and grew.

The making of the United States of America into a national state is even more revealing. The creation of the Continental Congress was the work of factions within the various colonies, factions with similar ideas of what ought to be done, factions which crossed the colonial boundaries in order to co-operate to get those things done. Later, the Articles of Confederation were adopted to give a better legal basis for the emerging nationalism. What gave the Articles the little actual vitality they possessed was the disposition of groups to cross state boundaries and to use the Congress under the Articles to group advantage. True, each state had but one vote in Congress, and under the unit rule that vote was cast as the majority of the state's delegation wanted it cast. But the vote of each individual delegate was tabulated, and the journal revealed that frequently a minority from Virginia was co-operating with a majority from New York or a minority from Pennsylvania was co-operating with a majority from South Carolina. Factional, group, class, and ideological differences were breaking through and crossing state lines.

The present Constitution of the United States represents the triumph of a class. The selection of delegates to the Philadelphia Convention of 1787, the activities of the Convention itself, the struggle for adoption in the various states, and the first ten years of implemental and constructive government under the Constitution were largely the work of the confident and aggressive commercial and creditor class which boldly crossed state lines to create a government which it thought would be advantageous to itself. And the groups opposed to the Constitution, also for reasons thought to be advantageous to their class interests, likewise crossed state lines to use within every state concerted tactics and arguments in opposition.

Since the adoption of the Constitution of 1787, nationalism in America has grown apace. The geographical and territorial conflict has given way still more before the class, group, and ideological conflict. The decisive divisions within the United States have not involved state against state or even section against section, but group against group. These group differences have cut across state and sectional barriers. The little states have not combined against the big states, but instead similar groups within the little states and the big states have combined against dissimilar groups within those states. Even sectional differences largely represent the preponderance of one group within a section as opposed to the preponderance of a different group within another section. The industrial minority in an agrarian area has usually co-operated with the industrial majority of another area just as the agrarian minority in an industrial area has usually co-operated with the agrarian majority in another area.

George Washington had scarcely been inaugurated as President when the Federalists even more than before were crossing state lines to co-operate with Federalists, and Republicans were crossing state lines to co-operate with Republicans. Thomas Jefferson had more in common with an Albert Gallatin of Pennsylvania and a John Langdon of New

Hampshire than he had with such fellow Virginians as George Washington and John Marshall. Likewise, Alexander Hamilton had more in common with the Essex Junto of Massachusetts and the Pinckneys of South Carolina than he had with fellow New Yorkers like Aaron Burr and George Clinton. And so it was with the political leaders and the rank and file in every state of the Union. The political conflict was not territorial; it was partisan, factional, class, group, and ideological, cutting across and breaking down state barriers.

And what had produced this accelerated trend toward group rather than state conflict, that is, this trend toward nationalism? In particular, the new government machinery helped. The states in Congress no longer voted as states, no longer voted as units. The votes of senators and representatives in Congress were counted as individual votes, and thus the group conflict which cut across state lines was facilitated and made manifest for all to see. Moreover, representatives in Congress were now chosen in popular elections, and thus a given delegation in the House represented a very wide cross-section of the state's opinion. (Under the Articles of Confederation state delegations in Congress had been chosen by the legislatures.) Representation of greater diversity shattered the state's political unity in Congress. More important, the new government had more functions which affected the everyday lives of individuals in their group and class relationships—functions and powers involving taxation, trade, tariffs, and monetary questions. Finally, the steady growth of technology (it is so easy to forget that the Industrial Revolution came to the world just about the time our Constitution was adopted) created new groups and classes, sharpened group and class consciousness, and caused government decisions to affect more closely economic and social matters.

The one big exception to this trend toward American nationalism—the growth of a separate nationalism in the South in the 1850s—underscores the importance to nationalism of class and group conflicts crossing state and sectional lines. When the Southern states in a solid political bloc faced the rest of the country, when Southern delegations in Congress no longer crossed the sectional line to vote with Northern congressmen, when party ties with the North snapped, when first the Whig and then the Democratic party split into Northern and Southern wings, when the groups and classes of the South no longer felt affinity with similar groups and classes of the North and ceased co-operating with them, then American nationalism sickened and nearly died. Nothing better illustrates the proposition that nationalism is based not on law or on government structure but on social feelings and realities. The Constitution of 1787 still existed in 1860, the sovereignty yielded by the states in 1789 still remained legally yielded in the 1850s and existed in the Washington government, the states still could divide their votes in Congress, social and economic groups and classes in the South were still free to co-operate with similar groups in other parts of the country—but all this availed nothing when a combination of social and economic forces converged to cause a whole section to think again largely in geographical and territorial terms rather than in group or class terms.

In short, local and territorial conflict gave way to political nationalism when group and class lines crossed local boundaries. It is probable that genuine political internationalism will come only when group and class lines reciprocally cross the old national boundaries.

III

Today an international organization based on contract, the United Nations, is struggling to become an effective world agency. Not much sovereignty, not much real power, has been yielded to it by the nations. People do not sufficiently feel the impact of the organization in their everyday lives. Consequently groups within countries do not cross national boundaries in political alliances with like groups abroad in an attempt to win hot political decisions within the United Nations—the sort of vital political activities that would infuse life into the international organization, break down nationalism, and develop real world government.

The United Nations does contain within itself the seeds—only the seeds—of the kind of power that would enlist the passions and practical interests of all groups and classes and ultimately force them into international alliances in a struggle for decisive political influence. Effective control over war and peace is one of these. The international trusteeship system is another. Matters dealt with by the Economic and Social Council suggest still other such powers. Most important of all is the American proposal to create an international agency with monopoly powers to manufacture and control atomic energy. Such an agency, having control over an increasingly important source of industrial energy, would come to influence directly the daily lives of every man, woman, and child on this planet, an influence that would be felt and seen constantly. This is the sort of political power that would impel economic and social groups in one country to make international alliances with like groups in other countries to affect the decisions of the control agency.

One of the most encouraging features of the United Nations is the open, spirited, and realistic nature of the debates in the Security Council and in the General Assembly. This is a distinct gain over the way debates were conducted in the Council of the old League of Nations. These debates sometimes become so heated that one suspects there must be division of opinion within the national delegations. But as presently organized there is no way for these divisions to make themselves manifest, no way divisions in one delegation can make contact with divisions in other delegations.

Indeed, this is a most serious lack in the United Nations—the inability of divisions within the national delegations to break through and beyond the national boundaries. There is no way a national delegation can divide and proclaim that division to the world in the way, for instance, that New York's delegation in the American Congress constantly divides on public questions. Each nation represented on the Security Council has but one delegate and one vote. Every member nation is represented in the Assembly by a plural delegation, but each nation has but one vote. There is no way of splitting that vote. Moreover, the votes of the individual delegates in the Assembly are not recorded as individual votes. The record does not show whether, for instance, on a given question France's five delegates in the Assembly voted three one way and two another. Perhaps this may have actually happened when the delegation caucused, the one vote of France, however, going under the unit rule the way the majority of the delegation voted. But if delegations do divide, there is no sure way of knowing it and no way for a minority in one national delegation to unite with minorities and majorities in other delegations who feel the same way about a particular issue.

It is apparent that the authors of the United Nations envisaged a society of sovereign states debating and voting in solid national blocs, each nation presenting a united front to the world. Nationalism might be "curbed" and some national sovereignty yielded, but in the main nationalism was to continue and national solidarity to be maintained. One was going to be allowed to eat his cake and have it too. Even our "Federalists" and our "Union Now" advocates talk too much in terms of "curbing" nationalism and yielding sovereignty—talk too much in sheer legalistic terms—and do not see the importance of functional alliances of groups and classes, minorities and majorities, crossing national lines and co-operating within a world organization for their own political ends.

The international organization which sought in its structure to encourage a functional division within the nations and within national delegations and functional co-operation of groups across national boundaries would provide for plural delegations and plural votes in the Assembly and perhaps even in the Security Council. Moreover, it would allot delegates and votes at least in the Assembly according to a nation's population, technology, and capacity to produce. And it would provide for the popular election of the delegates, to insure their being representative of a wide cross-section of opinion and of all articulate groups and classes. Now in the present state of the world it is too much to expect that all of this can be accomplished. For a time we shall have to maintain equality of delegate and voting strength even in the Assembly. But at the very least we could in the Assembly call the roll not of the states but of the individual delegates and thus proclaim to the world divisions within the delegations, although under the unit rule the one vote of the state would, of course, go the way the majority of the delegation voted. As we have seen, even our old Articles of Confederation went this far! And we might go to the length of providing that the national delegations in the Assembly be chosen by their respective national legislatures, not their executives. This probably would provide a somewhat wider national cross-section and be more inclusively representative of the various classes and groups.

IV

Are not the forces of nationalism too strong to prevent groups and classes within nations from crossing their respective national frontiers and co-operating with similar groups abroad? Would such forces operate internationally even if every encouragement were given them by the legal and mechanical structure of the world organization?

As a matter of fact, ideological co-operation on an international scale is rather common in history and occurs today. During the American Revolution, British Whigs sympathized with American Whigs and found opportunities to lend them practical encouragement. During the Wars of the French Revolution, American Federalists openly sympathized with British Tories and American Jeffersonians with French Girondists and even Jacobins; and while the War of 1812 was in progress many American Federalists were at bottom in sympathy with British Tories. During those same French Revolutionary Wars, while Britain and France were locked in deadly combat, the Charles James Fox Whigs went to very great lengths to show their sympathy for the French Revolutionaries. At the same time, British and Continental conservatives were co-operating with French conservatives in opposition to the Revolution.

Such international ideological co-operation has become much more marked in the twentieth century. Today it is a commonplace for Communists of one country to co-operate with Communists of other countries, for the fascist-minded of one country to co-operate with the fascist-minded of other countries, for Socialists of one country to co-operate with Socialists of other countries. The Second World War was not merely an international war; in many ways it was converted into a gigantic civil war with Communists and Socialists in the Axis countries doing all they could to defeat their own countries and with the fascist-oriented in all countries furthering the cause of the Axis, even at the expense of their own countries. Today left-wingers in various countries co-operate to discredit Franco Spain; conservatives in various countries co-operate to sustain Franco. Much the same ideological conflict cuts across national lines with respect to Indonesia, French Indo-China, Argentina. When Winston Churchill speaks to Americans on questions of American foreign policy in a way that offends millions of left-wing Britons and when Henry Wallace goes to Britain and lashes out at American foreign policy in a way that pains conservative and middle-of-the-road Americans but pleases left-wing Americans, we are witnessing at first hand and in a conspicuous way this increasing tendency of ideological groups inside countries to cross national barriers and cooperate with like ideological groups abroad. Of course, this sort of thing produces violent protests from nationalists and patrioteers—if it involves liberals and leftists co-operating internationally—but the tendency persists and grows.

This tendency increasingly manifests itself in a growing number of functional organizations. Among these are: the Deminform (promoted by the AFL); the Cominform; international business organizations and cartels; international trade, financial, professional, scientific, literary, religious, and youth conferences; world-wide gatherings of educators under the sponsorship of UNESCO; the meeting of business and labor leaders from many countries in the General Conference of the International Labor Organization; the commingling of Communist and non-Communist labor leaders in the WFTU and the banner of free unionism carried by James B. Carey into that organization; the international co-operation of the trade unions of the sixteen countries which are participating in the Marshall Plan and the insistence by these unions that labor be adequately represented in the administration of the program.

Now if this trend is as observable as it is today, how much more observable it would be if we had a world organization with real powers—powers which affected the practical lives of all people, all groups and classes, in their bread-and-butter daily existence. A world organization with broader economic powers would impel manufacturers, bankers, traders, exporters, importers, technicians, engineers, planters, farmers, wage-earners, white-collar employees, professional groups, intellectuals, and consumers to cross national lines and to join hands with kindred groups in other countries, each group seeking to influence and control the international organization in its own interests. In short, the same process of group conflict and compromise which now characterizes national politics would be extended to international politics. For a long time, of course, the most numerous and most important conflicts would still be within the nations, especially in this age of planned economies, and even in Communist countries the chief questions would continue to be resolved within national frameworks, but gradually the area of international group conflict and compromise would widen as the functions of the international organization widened.

V

But now we must pause to take note of the most serious obstacle in this mid-twentieth century to the building of real world government. It involves this very trend we have been discussing. It involves the unilateral, nonreciprocal way the Communists use ideological international co-operation. At first sight it would seem that the high degree of co-operation of Communists from country to country would be a promising indication of the early arrival of world government—indeed, the most promising indication in the world today. But the rub is that the Communists insist that international ideological co-operation be a one-way street. Communists in Communist-dominated countries insist on being allowed to co-operate with Communists in non-Communist countries and do so co-operate. But non-Communist elements in non-Communist countries are not allowed to co-operate with non-Communist elements in Communist-dominated countries. Russia can cross national frontiers and co-operate with Communists in Western countries, but non-Communist elements cannot cross the Soviet frontiers and co-operate with whatever non-Communist elements exist there. And, using this same principle in reverse to designate ideological opponents abroad, a Vishinsky can go inside the American delegation to the Assembly of the United Nations, seek to stir up factional division there, and point the finger of scorn at one sitting with the American delegation, but a Marshall cannot do the same to the Soviet delegation, which, representing a totalitarian state, always presents a solid, an impersonal, face to the world.

When international ideological co-operation is not reciprocal for all groups in all countries, there is danger that those countries which do not allow ideological division inside their own countries will use ideological division within other countries in a fifth-column and Trojan horse manner to advance national ends or to create international dictatorship. Thus the very people, the Communists, who boast the greatest degree of international ideological co-operation among themselves, are staying the larger trend toward international ideological co-operation or are forcing it unnaturally into two clear-cut ideological camps which threaten ultimate war.

It is possible, even probable, that the Communists will not be able forever to prevent new group conflicts from emerging openly inside Communist systems—the political conflict of technicians, engineers, managers, wage-earners, farmers, and so forth. For even Communist purges will not permanently be able to stifle the development and conflict of new economic group interests. Those group conflicts will eventually spill over national boundaries and seek to co-operate with the nearest thing to similar groups abroad. And believers in the democratic way should take every opportunity to make contacts inside Communist countries and to encourage democratic divisions, divisions that may cross national borders. Only about a decade ago dissident groups and leaders were active in Russia. It required a ruthless government purge to silence them, and opposition remnants no doubt remain. That the Voice of America is making some impression is evidenced by the vigorous protest which came out of Moscow recently, a protest which filled nearly three columns in the New York *Times*. The face-to-face plea for a free press made to Soviet journalists by Melvin J. Lasky, a correspondent of *The New Leader*, at the Soviet-sponsored German writers' conference in Berlin last fall, is the sort of realistic encounter that may bear eventual fruit. International youth congresses in which Communists and non-Communists from many nations mingle together should be encouraged by our government rather than discouraged. (Democratic youth can be counted on to hold its own against the Communists.) And the mixing together of Communist and non-Communist labor leaders from many countries in the WFTU is all to the good. However, in the present stage of Communist development such contacts are not too promising. Communist tactics today, by insisting that international ideological co-operation be a one-way street, are slowing up and may even defeat the evolution of world government based upon consent.

VI

However, we may be nearer to international or supranational government than we suspect. Many forces—technological, ideological, psychological—are carrying us in that direction. But the *kind* of international government we develop is as important as the fact of international government itself.

One possibility is that the current duel between the United States and the Soviet Union would ultimately lead to war and that the victorious survivor would emerge with sufficient strength to put the world together by force, veiled perhaps, but force nevertheless, and thus create a universal state dominated by one power—a modern *pax romana*.

Another is that the Communists, employing international functional co-operation, would come to power in enough countries to create the international proletarian state—the world Communist dictatorship—after the manner contemplated by the old Comintern. (However, even the Communist systems might turn out to be more nationalistic than Communist theory had planned.) In the long process of history, new articulate groups would emerge within the Communist systems and the resulting political conflict would probably lead away from world dictatorship to a more democratic world organization, but that is speculative and for the remote future. The immediate result probably would be international Communist dictatorship.[1]

A third possibility is that a world organization founded on contract —the United Nations, for instance—would evolve functionally into a genuine world government. This may come about if the tensions between the Soviet Union and the United States recede—after the respective spheres of the two countries are delimited. Then more power might be granted the world organization or might gradually develop from the

1. The previous article, "The New Nationalism," written later than this one, pointed out that by 1950 these first two possibilities had in all probability already miscarried.

34

powers already granted. Also delegations might be allowed to split their votes and the functional co-operation of groups that crosses national boundaries and gradually disintegrates nationalism be encouraged. However, the great barrier to the breaking down of nationalism in this manner would still be the fact that delegations from democratic countries would be free to split their votes while delegations from dictatorship countries would not, even though legally allowed to do so. It cannot be stated too often that the functional development of group politics on an international scale must be reciprocal, must be a two-way street. The easing of tensions, and splits in the delegations from Western countries might induce Communist countries to relax dictatorship and allow open political differences, differences which reflected themselves in split Communist delegations; but that, of course, is extremely doubtful. Indeed, the doubt is so great that it would probably prevent Western delegations from splitting in the first instances. Thus, in the presence of dictatorship, the process probably would never get started.

A fourth possibility is that the Soviet Union and its satellites, distressed at repeated rebuffs within the United Nations, would secede from that organization, thus giving the Western democracies an opportunity to strengthen the United Nations at a more rapid pace—by granting it more powers or developing more powers from the ones already granted and encouraging the functional co-operation of groups that cuts across national lines. Admittedly, there are dangers here. One is the overwhelming power the United States might exert within the smaller organization. Another is the fearful possibility that the United States, with its doctrinaire belief in the righteousness of free enterprise and pure capitalism, might use that power to prevent the synthesis of capitalism and democratic socialism in Western Europe—the synthesis which stands the best chance of checking Communism there. Of course, these dangers exist even today, but they might be increased should the United Nations shrink in size by the secession of the Soviet Union. However, if the United Nations is revamped to allow national delegations to split and if national delegations actually do split, then leftists in the United States probably would join hands with the Socialists of Britain and Western Europe to check possible American domination or American smothering of the developing socialist-capitalist systems of Western Europe. It is well to remember that in the early days of our own American federal system the fears of the small states that the large states would dominate the Union proved to be groundless precisely because the political conflict developed, not as state against state, but instead as group against group, party against party, which conflict cut across state lines. This is the way world federalism would operate if it were to operate successfully.

Still another danger is that Soviet secession from the United Nations, by severing the world into two hostile parts, would further increase world tensions. This might well happen and lead to a situation which would make impossible any effective, democratic international organization in our time. On the other hand, the added strength that would come to the non-Soviet world through the establishment of a workable federal state might actually ease tensions by mitigating fears in the Western nations. And, perhaps most important of all, such a federal state, composed of Western democracies, would provide a developing method for the peaceful and democratic synthesis of the "isms" now contending for supremacy in the Western world—capitalism, socialism, and Communism. For, as James Madison pointed out in his most famous essay in *The Federalist*, the wider the territory and the more people and groups involved in internal governmental conflict, the less danger there is of the triumph of any one extreme, the greater the chance for broad, middle-of-the-road compromise.

The evolution of the United Nations into international or supra-national federal government is today an alternative to a world state based upon militarism or upon Communist dictatorship. It alone offers international government based on consent. It alone combines strength and democracy. But we must constantly bear in mind that federalism means more than mere legal structure; it means more than the surrender of a measure of national sovereignty. *It is far more a matter of politics than of law.* If international federal government is to come to life, it must be infused with the vitality that only the competition and co-operation of functional groups, bursting national boundaries, can give it. This is a basic truth our professional "federalists" too often ignore.

❧ An Atlantic Curtain?

The following article appeared while McCarthyism was at its height, and it examines the intolerance of that time. The essay describes the alarm in Europe over the hysterical state of opinion in America (a description derived from the author's observations while traveling in Europe), and it recalls how even many of the liberal elements in America, while deploring McCarthy and his tactics, were themselves affected by the conditions which produced McCarthyism and showed little understanding of the strictures on American foreign policy then prevalent in Europe. Even so moderate a periodical as the *Saturday Review*, in its July 18, 1953, issue, editorially pronounced the following article as "ominous," and it saw no value in harping on European criticism of America and its foreign policy.

Rereading the article after the passing of the years, the author was struck by the way we today have come to have a mistaken interpretation of McCarthyism, the strange degree to which we personalize it, give McCarthy himself too great a play, ignore the conditions out of which McCarthyism emerged. The fall of China to the Communists had taken place only a few years before; Americans were still agitated by the Korean war; the United States was new to the world leadership and hypersensitive to criticism; Secretary of State John Foster Dulles was urging on our allies Cold War policies which they considered extreme; and at that time he was talking about "a rollback" of the Communists in Eastern Europe, castigating neutralism as "immoral," encircling the Communist powers with containment military alliances, and proclaiming the doctrine of "massive retaliation" by atomic weapons.

An element of McCarthyism has survived to this late date. A distorted view of Communist solidarity lured us into the Vietnam war, and much of the world's criticism of that war is reminiscent of its criticism of American policy at the time of McCarthyism. Only now, in 1970, are Americans coming to feel that perhaps the world situation was never as inflexible as they had supposed and that a monolithic international Communism, if not a myth, was certainly highly exaggerated.

The American Scholar, XXII (Summer 1953).

■ INFLUENTIAL EUROPEANS have two misgivings with respect to American leadership in world affairs. They fear that Americans are hysterical and war-minded and in process of imposing thought-controls on their fellow Americans; and they feel that Americans still do not understand European conditions and attitudes. As a matter of fact, these sentiments are linked together and grow out of the same situation.

Several times recently Bertrand Russell has stated publicly that liberty is drying up in the United States, that the only places where real freedom any longer exists are in western Europe, particularly in Britain and Scandinavia. These statements shock Americans, but they are regarded as true by many leading Europeans. Europeans often say, "If the Americans had to deal with the Red Dean of Canterbury, they long ago would have defrocked him and sent him to jail. The English merely listen in irritated and tolerant amusement." Not long ago an official organ of the Anglican Church ended a ringing editorial on the need to maintain freedom with this sentence: "Let us be thankful that in Britain we have nothing so un-British as a committee on un-British activities."

Recently an exchange graduate student from Germany called my attention to an issue of an American magazine of wide circulation which carried an article entitled "Confessions of an American Liberal." Referring to this article, the German student asked, "Is it not enough that you Americans demand recantation from Communists and even Socialists? Do you now demand it of all liberals?" This same German student later asked, "What do you Americans mean by the word 'egghead'? Is that some kind of intellectual? Are you Americans ridiculing all creative and talented people, all people with ideas? Has it become a sin in America to have an idea?"

Europeans point to the "witch hunts" staged by congressional committees; the proliferation of "totalitarian" demagogues like McCarthy and Jenner; the tighter security guarantees surrounding and inhibiting American scientists; the imprisoning of American Communist leaders, not for overtly hostile acts, but for constructively hostile ones. (The French say that they, unlike the Americans, caught their Communist leaders actually directing riots but that they released them.) Europeans also have taken note of the suppression of that sensitive movie *The Miracle*; the threat of the Department of Justice to bar Charlie Chaplin's re-entry into the United States; the refusal of the State Department to permit Alberto Moravia to visit this country; the barring of playwright George Kaufman from television for a time because of an "irreligious" remark; the *arrest* of a heckler who put a question to Senator McCarthy while the latter was speaking to an electioneering banquet last October; the debate at a national meeting of secondary teachers of the social sciences at Dallas last autumn in which the very word "social" in social sciences had to be defended from subversive connotations; the increasing suspicion and condemnation of schools, teachers, and textbooks in the United States.

Many Europeans place together bits of evidence like these and conclude that a veritable reign of terror is at work in American schools and colleges, that a grand inquisition is in progress against all Americans who deviate from narrowly accepted thought.

There is some truth in the European observations about America's growing thought-control, but the thought-control that exists in America is taking a somewhat different turn from that conceived by Europeans. There is no overt reign of terror among our intellectuals. Relatively few teachers, professors, editors, and writers have actually been dismissed from their positions, although Hollywood, radio, and television artists have increasing difficulty getting "cleared." Individual intellectuals may still be found who deviate from current stereotyped thinking. These individuals usually keep their jobs. However—and this is the nub of the matter—nonconforming individuals have less and less influence. Fewer and fewer avenues of expression are open to them. Fewer and fewer people take them seriously.

The greatest danger to freedom in America does not consist in espionage and personal persecution; the greatest danger is that a growing consensus, an omnivorous stereotyping of thought, is drying up the sources of diversity and deviation and intellectually isolating all those who do not conform. A nongovernmental, self-imposed totalitarianism in ideas seems to be creeping over American life. The danger is not terrorization; the danger is that there will be few individuals to terrorize. The danger is deadly intellectual inbreeding, and consequent intellectual isolation from the thought of the rest of the world.

Western Europe, although physically closer to the Communist menace, has escaped so complete a uniformity of thought. Europeans have lived with the Communist threat since 1917. In some countries, like Italy and France, non-Communists in their everyday activities are compelled to associate with many actual Communists—in stores, in restaurants, in theaters, at dances, and at social affairs generally. Communists are not men from Mars. They are not thought of wholly as devils, and the Communist movement is not conceived of as wholly irrational and foolish. Most Europeans realize that Communism springs in large measure from conditions and that conditions can best combat it.

Again, the political parties of European countries are more numerous, more varied, and the differences among them more sharply drawn. Western European countries have capitalist parties, but they also have Socialist and social-democratic parties. These latter parties do not view events from capitalist points of view. Even in Britain, where in the main a two-party system exists, the parties are wider apart ideologically than American parties. These more widely diverse parties of Europe act as mass agencies to keep their national publics conscious of multiple possibilities, alternatives, and options in foreign and domestic policy.

Periodicals in Europe reflect this wide range of opinion. A single issue of a periodical on the Continent or in Britain may contain an article by a Louis Aragon, a Saragat, a Silone, a Croce, a Maritain, a Sartre, a Malraux, a Russell, a Morrison, a Crossman, a Bevan, a Harold Macmillan, and a T. S. Eliot—or, more commonly, a galaxy of lesser names expressing a similarly diverse range of views and feelings. And while there are relatively few individual periodicals with such catholicity of scope, every point of view represented by the names just listed has open to it important periodicals.

Moreover, the better European periodicals carry a number of "heavy" articles which contain a large amount of factual information combined with mature and reflective interpretation. When a magazine like *Twentieth Century* or the *Economist* does a piece on Thailand or Indonesia or Iraq or Turkey or Sweden or Argentina, the article is not written with the question uppermost: Are these people for us or against us? Are they for the "free world" or for the "slave world"? Rather, the article is likely to be written from the point of view of the country under discussion and to deal in rather full and almost always realistic fashion with the social classes and political groups in conflict in that country.

American society does not possess the forces and agencies which make it easy to maintain a wide range of divergent thought in the face of grave international crisis. There are only two vital political parties in America. No third party offering a significant new approach any longer exists, and the choice between the two major parties is relatively narrow. Both these parties, because of their incongruous and widely inclusive natures, gravitate toward the center. Political differences are blurred and obscured. The bipartisan foreign policy has narrowed further the differences between the parties and has resulted in putting a far larger emphasis in American foreign policy on armaments than on undermining Communism by positive programs of popular mass appeal and basic social reform in Europe and Asia. This preponderant emphasis on military solutions has strengthened the conservatives of both parties.

America's "underprivileged" groups—the wage-earners, the share-croppers, and the Negroes—are so "privileged" by European and Asiatic standards that their point of view is likely to be as conformist and nationalist as that of other American groups. Our wage-earner and Negro groups share so much of America's national prosperity that they scarcely function as links to Socialist and social-democratic parties abroad or as vehicles for popularizing in America the attitudes of Socialist and social-democratic parties in Europe. The article contributed by Walter Reuther to the "Third World War" issue of *Collier's* shows how widely the thought of a representative "radical" labor leader of America differs from the thought of European labor.

American periodicals reflect American political uniformity. An American liberal periodical, the *New Republic*, attacks Britain's *New Statesman and Nation*, a leading left-wing Socialist periodical, as disloyal to the free world. The *New Leader*, an American periodical of Socialist origin, is much closer in point of view to the *Freeman*, that four-square exponent of American free enterprise, than it is to the socialist periodicals of Europe, be they left-wing, right-wing, or moderate. Indeed, the *New Leader* and the *Freeman* have many contributors in common. The *Nation*, which consistently reports to Americans the point of view of Socialist and social-democratic parties in Europe, is now almost constantly under attack by American liberals.

When an American periodical of large circulation carries a story on a European, an Asiatic, an African, or a South American country, that story almost invariably approaches the material from the *American* point of view. Is that country on *our* side? Has it ranged itself squarely for the free world, that is, on the side of American foreign policy? If it has, the article is likely to be accompanied by photographs of peasants in neat native garb; of that country's new army with its shiny, up-to-date equipment furnished by American money; of modern buildings in that country's capital city, buildings likewise made possible by American money. Rarely, almost never, does the story deal in realistic fashion with the actual group and party conflicts taking place in that country.

The growing uniformity of American thought is making it increasingly difficult for Americans to learn and understand the views of their allies in western Europe. A small and shrinking segment of European opinion shares America's middle-class attitudes and sympathizes with American free enterprise ideas. But the overwhelming mass of people in Europe are more socialist in their thinking—even those who do not belong to socialist or social-democratic parties, even those who belong to the Catholic parties of the center. The repeated disasters which have befallen Europeans since 1914 have largely destroyed the middle classes and middle-class attitudes.

A prominent Frankfurt editor, who believes strongly that Communism must be contained and that America and western Europe must work together to contain it, recently visited America. He was amazed at the appalling gulf between shellshocked Europe and the cheerful hedonism of Americans "with their great overriding urge to make the world, or at least America, safe for middle-class comfort and leisureliness." He found Americans frighteningly remote from the poverty-stricken and sullen atmosphere of the Old World.

Another European editor, recently in America, likewise a friend of American foreign policy, expressed himself as feeling that there was as much difference between America and western Europe as between western Europe and eastern Europe. "Are the Americans ringing down an iron curtain in the Atlantic, cutting themselves off from understanding their allies in western Europe?" he asked. The thought barriers separating Europeans and Americans hardly constitute an *iron* curtain or even a curtain—perhaps a screen better describes them—but what is needed are bridges of communication and understanding, not barriers.

Most articles about European opinion of things American which appear in American periodicals are misleading; they are expressly written with a view to avoid shocking or wounding Americans. For instance, it is doubtful if even the articles of D. W. Brogan and Barbara Ward, which not infrequently appear in the Sunday New York *Times Magazine*, would be publishable in England as a representative attitude of the British toward American policies. Sometimes Bertrand Russell, peppery old rascal that he is, publishes articles in America about British and European opinion of things American that are candid enough to be published in Europe too. Some of the articles of Alexander Werth which appear in the *Nation* also appear in European periodicals. But Russell and Werth are exceptions.

Few Europeans, except doctrinaire Marxists, think of America as a fascist country. The analogy most often made by informed Europeans is that which compares modern Americans to ancient Romans. Americans are thought of as doers and organizers from a wealthy land, possessing little maturity or sensitivity in things not immediately practical. (Are Americans really interested in anything except applied technology and their own precious standard of living? Europeans often ask.) This analogy to the Romans is flattering to Europeans, for they think of themselves as Hellenes who must teach Americans greater maturity and better acquaint them with the rich diversity of the Western heritage.

But many Europeans have reservations about the Americans becoming the successful Romans of the modern age. Will the Americans come to know enough about the vast non-Soviet world in time to organize it effectively? Can they develop sufficient political maturity in time to lead so large a part of the world? Will they be tolerant of the wide and rich diversities existing in this far-flung world, sufficiently imaginative and sympathetic in their approach to its many and various cultures and points of view? As Americans more and more insulate themselves from European points of view, increasingly and defensively identifying the good life and the salvation of Western civilization with economic free enterprize narrowly conceived, Europeans wonder.

Not long ago I had a heart-to-heart talk with a well-known professor in an Italian university. He made a comment about Americans something like the following: "Americans were once an earthy people suffering the vicissitudes of the rest of mankind—in the wilderness and along the frontier, even more than their share—and having a sense of individual and of collective tragedy. From earliest colonial times they were also a highly diverse people, containing within themselves many different nationalities, religions, and subcultures. All of this made for an affirmation of qualitative human and social values that linked Americans with the deepest emotions of Western civilization and of mankind. But at home Americans have not experienced profound emotional tragedy of a collective sort since the Civil War (although since that time most of the rest of mankind has undergone collective sufferings of the most poignant kind); and through the widespread use of machines Americans have removed individual tragedy a few degrees, disguised it, and given it a new dimension. Americans have mechanized their emotions and standardized their ideas so that these have become mere popular stereotypes, mere quantitative reflections of mass wishful thinking, mass illusions, and mass half-truths. This has resulted in superficiality, in emotional thinness, and in a kind of narcissistic self-adulation. When Americans project their homemade stereotypes into world affairs, in which historically they have had little experience, the result is even more serious distortion. Moreover, America's national leaders have distressingly little humanistic and historical background. They seem to be mostly businessmen, business lawyers, advertising men, military men, and narrow specialists and technicians."

When Europeans ask that Americans become less stereotyped in their international thinking, more aware of cultural diversities in the world, and more alive to the alternatives in the world situation today, what do they mean? Let us consider these in more concrete fashion.

First, Europeans feel that the Americans do not understand or will not admit the forces making for greater collectivism in Europe and Asia. And they are frightened when Americans fail or refuse to make the very important distinction between Communists and Socialists. The mass of western Europeans are socialists of a sort, whether they actually belong to a Socialist party or not. They say that if Americans cannot learn to live gracefully with socialist allies, the Grand Alliance ultimately will founder. To Europeans, American public men who put Socialists and Communists in the same boat and who publicly denounce socialism are potentially as isolationist as those super-nationalistic Americans who today are denouncing the United Nations and UNESCO.

Second, many Europeans believe that the Americans do not understand the nature of the revolutions now taking place among the "backward" peoples. Almost all informed Europeans see these revolutions as partially combinations of native nationalism (or "resurgent culturalism," to use the term of F. S. C. Northrop) and indigenous collectivism, and not, even in China, as mere outthrusts of Soviet Communism. Collectivism often arises from the urge of "backward" peoples to move rapidly from feudalism to industrialism, to acquire modern and expensive technology in a hurry and thus drastically raise their living standards. Since these backward peoples have no extensive native middle classes and capitalist classes, they must provide the capital for widespread and rapid industrial development either by borrowing from foreign capitalists (and seeing much of the benefit siphoned off by American and European capitalists) or by creating capital through collective government action. The backward peoples are choosing the latter course as apparently more advantageous to themselves. Therefore all contemporary nativist revolutions of national liberation also contain large elements of collectivism. Sometimes that collectivism is extreme, as in Communist China; sometimes it is milder and seeks to go slower and to develop both bread and freedom simultaneously, as in the social democracy of Nehru's followers in India. But the native collectivism is always there, differing in degree from country to country. An English journalist once said to me, "In the days of the French Revolution and in the days of the revolutions of 1848, when revolutions were moving in the direction of American experience, Americans had a poorer press but a better understanding of the world situation than they have today."

Third, informed Europeans are convinced that if Americans really understood the reasons for and the depth of the collectivist movements among backward peoples, they would take their Point Four program much more seriously, even to the extent of elevating it to an equal place with their armament program. Mere armaments, Europeans feel, will never win the ideological conflict. Only a program which drastically raises living standards will do that. Indeed, a program which puts its overwhelming emphasis on armaments is likely to leave too little capital for an effective Point Four program and to depress seriously the living standards of America's European allies. Even should the Western world win a war against the Soviet Union, it still would be confronted with what to do about the native nationalist and indigenous collectivist revolutions of the colonial peoples. Americans, then, may be thinking too much in terms of their direct relations with the Soviet Union and not enough in terms of a flank attack on the Communists, not enough in terms of keeping their allies and winning the undecided peoples. Arnold Toynbee says that the balance of power is to be won or lost in the areas still uncommitted. America's recent rejection of the United Nations Covenant on Human Rights certainly will not help us in the battle for uncommitted minds.

Fourth, Europeans think that the Americans are failing to see sufficiently the strong nationalist element even in the Communist revolutions. The Communist world revolution has turned out in fact to be a series of Communist national revolutions, taking place in different backward countries at different times, and for the most part building industrial revolutions along national patterns in those countries. The form of Communist organization and the aims of the Communist movement in China are not identical with those of Russia. Moreover, the Chinese do not want the Russians in Manchuria or at Port Arthur or Dairen any more than they want other foreigners there. The Chinese revolution was largely a native, indigenous revolution. (However, it was no mere liberal agrarian revolution!) Help from Russia was late and small. Indeed, at this advanced date, Russia could not put a stop to revolutionary movements and activities in Asia, even Communist ones, assuming Soviet leaders were of a mind to do so. In Europe, even conservative leaders and journals do not believe that China is an abject satellite of Russia. Almost all European leaders, left, right, and center, believe that national differences between China and Russia and between some other Communist countries and Russia will take place in time. (Tito's Yugoslavia is an early and extreme expression of what is likely to come, they believe.) They feel that the Western allies should encourage and exploit national differences and seek to substitute a multiple balance of power for the dangerous polarization of power now existing.

Yet in America it is the rankest heresy to suggest that China is not completely dominated by Russia, that every Communist advance anywhere in the world (except in Yugoslavia) is not a direct advance of Soviet imperialism, is not necessarily another step in the dangerous upsetting of the world balance of power. During the height of the presidential campaign of 1952, Walter Millis asserted in the New York *Herald Tribune* that for either presidential candidate to hint that Titoism offered some hope of a way out of the international impasse, of a possible way to a multiple and less dangerous balance of power, would be to court political disaster. Unfortunately, Millis was emphatically correct. There is simply no place in the present American stereotype for such an eventuality.

The day after the popular victory of the Ibáñez presidential candidacy in Chile, the New York *Times* carried an editorial pointing out the extreme Chilean nationalism in the Ibáñez position and deploring the sweep of two destructive forces in the world: communism and nationalism. The editorial spoke of the nationalist revolutions in Asia, Africa, and South America, mentioning among these the Chinese revolution. Yet apparently it never occurred to the writer of this editorial that the nationalism in the Chinese Communist revolution and in other nationalist revolutions might be used constructively to make Communist revolutions less destructive to the interests of America and the Western world. This most certainly would have occurred to editorial writers of that conservative old Rock of Gibraltar, the London *Times*. This difference in approach of the New York *Times* and the London *Times* well illustrates the difference in approach of conservative Americans and conservative Englishmen.

If Americans were more experienced in international affairs, greater consideration would be given to the possibility that historical forces ultimately might work us out of the present stalemate. (This has often happened in the past, more often than is realized, for people remember the crises solved by war much more vividly than the crises solved by underlying historical processes. In the half-century after Waterloo, for instance, Russia, Austria, and Prussia were in ideological alliance against democratic countries and posed a serious menace; but this alliance ultimately was split asunder by internal conflicts of national interest.) Americans, however, know too little of long-term historical processes and the long-run working of the balance of power; they are too much impressed with the Hitlerian expansion, their one clear-cut experience with the balance of power, which they read into the present crisis; they are too impatient of temporary frustration; they are too confident of their own wealth and strength; and they are too conscious of their own brief history of uninterrupted success and victory, the success and victory of virtue over evil, to be much impressed by underlying historical forces which ultimately and peacefully may restore a multiple balance of power. (At least this is the current European appraisal and fear of the American attitude.)

True, there are still imperialists in Europe who would cling to the old colonialism. There are Dutchmen who lay their country's present burdens to the Dutch withdrawal from Indonesia. There are Englishmen who would hold on to investments in Iran and Egypt, Frenchmen who would hold on to Tunisia. But these are dwindling attitudes; they are much stronger in financial and governing circles than among the European peoples. Moreover, it is not an easy thing to liquidate colonial investments of decades and even of centuries, no matter how clearly the contemporary forces of Asian and African nationalism are perceived. Still, it is remarkable with what grace the British have abandoned their vast empires in India and Burma and are making concessions to colonial nationalism elsewhere. Even in Iran and Egypt there is a steady yielding. And it is notable that even among members of the British Conservative party there is strong support for the Colombo Program, Britain's Point Four program in Asia, and for admitting Communist China, which has been officially recognized by Britain, into the United Nations as a means of encouraging Chinese independence in foreign affairs. French opinion is far more anti-imperialist than appears from French government policy, for it must be borne in mind that the basic split between French Communists and Socialists, who see eye to eye on questions of imperialism, makes government policy far more conservative on these matters than French opinion; and the weight of the investing interests inside the so-called Radical Socialist party and inside the Popular Republican movement is usually stronger in molding government policy than in carrying conviction in colonial matters to the rank and file of party members.

Fifth, Europeans distrust the American tendency to approach foreign policy in moralistic and legalistic fashion. They fear that Americans do not appraise in a realistic way the long-term ideological and balance-of-power considerations, that little by little the Americans are moving from policies of containment and security to a policy of eradicating Communism everywhere in the world simply because in the American view Communism is a moral evil. Europeans believe that if international politics is conceived as a conflict of moralities, then there never can be any basis for compromise or peaceful living together. International relations then become a conflict of irrationalities, a conflict of intransigencies, a conflict of fanaticisms. Why is it, asks the European, that the conflicts of the historical past almost invariably seem to be conflicts of half-good men (on both sides) and of half-good causes (on both sides), but that to Americans, more than to western Europeans, conflicts in the present seem always to appear as conflicts between white and black, good and evil? Again, Europeans are puzzled by the American habit of using lofty moral terms in foreign affairs while at the same time using those same foreign affairs as vehicles for getting party votes from national minorities in the United States in electioneering campaigns. The fact that the highest moral appeals (the moral appeal of a free Israel, the moral appeal of liberating Poles and Czechs from Soviet tyranny, for example) always accompany these campaigns for minority votes intensifies the danger in European eyes. For it is believed that both the moralistic appeal and the parochial appeal are dangerous, and when combined become doubly so.

Sixth, Europeans point out that Americans still continue along the path of economic isolation. Americans regard inflation and deflation in the United States as exclusively domestic matters. Although the United States is the leading creditor nation, Americans refuse to buy European goods and refuse even to discuss a genuine reduction of their tariffs and the economic integration of the North Atlantic community. Again, American consumption of automobiles, radios, television, electrical equipment, and numerous other mechanical gadgets takes a tremendous toll of the world's resources, but Americans decline to face the economic consequences for the world's future. In short, Europeans view the American disinclination to think seriously of economic planning on a world scale as a cardinal American "irresponsibility."

Finally, Europeans are uneasy about the increasingly nationalistic spirit in the United States. As American thought grows more uniform, it grows more nationalistic; and as it grows more nationalistic, it grows more uniform. Europeans see the United Nations and UNESCO under increasing attack in the United States. They witness the American debate over priority to Asia or Europe; they read statements indicating that more and more prominent Americans feel the United States is so strong that it could maintain a unilateral position in world affairs, that it could go it alone in isolation or in a kind of imperialism which browbeats its allies. The Korean war has intensified American nationalism, and the problems growing out of that war are increasingly divisive for the West, which may be the main reason the Communists refuse an armistice and keep the war going. The Republican victory of last November was widely interpreted in Europe as a victory for conservative and perhaps even unilateral American nationalism. The first moves of the Eisenhower administration in foreign affairs raised fears in Europe that American policies in eastern Europe and in Asia would take European opinion into consideration even less than postwar American policy has done in the past and might even be formulated in opposition to and in defiance of America's allies. Europeans ask: "Of what value is a dragooned European unity, of a Europe dragged along into policies against its better judgment, or of a unity that restores German power to the extent that it produces fear and anxiety in the rest of Europe? And of what good is a European army if it is not backed by an affirmative popular morale? European unity will come; but to be worth anything, it must come through a genuine meeting of minds."

Why has not the American press done a better job of informing Americans of opinion in western Europe? (For instance, Americans had to wait until General MacArthur was actually dismissed to discover the almost universal and profound distrust of MacArthur by all segments of European opinion.) In part, because Americans are too self-conscious about the world crisis and have an exaggerated notion of national security requirements, a notion which may go so far as to defeat some of its own ends; more important, because of the narrow stereotyping of public opinion in America, discussed earlier. The American press faithfully reflects the uniformity of American opinion, the lack of diversity in American classes, groups, and political parties.

In the eighteenth and nineteenth centuries, the most remarkable quality of British world leadership consisted not in Britain's extensive subsidies to her allies, not in her willingness to buy the goods of her debtor allies—

important as these were—but rather in her sense of vicariousness, her tolerance, her wide and agnostic inclusiveness, her hospitality to various points of view, to even the heresies and heretics of all nations. An illuminating passage in Stendhal's autobiography, *The Life of Henri Brulard*, illustrates how thoroughly this was understood, even by Britain's enemies. In spite of his adoration of Napoleon, Stendhal reveals how passionately he wanted Napoleon to *fail* to conquer Britain, for "where could one have taken refuge then?"

In this day of American world leadership, however, even international-minded Americans demand uniformity of thought. In hortatory speeches and articles they constantly exhort us to develop an all-embracing mystique that will unify American and European intellectuals and form a spiritual bridge for the Grand Alliance of the Western world. This is the Platonic ideal, which has come near realization only in authoritarian systems. This is no ideal for a free world.

When have the intellectuals of a single country of the Western world been united, let alone the intellectuals of the whole Western world? In the days of the Renaissance? In the days of the Reformation? During the Enlightenment? In the nineteenth century? Even a penetrating study of the Middle Ages will reveal as much intellectual diversity as unity. The truth is that the Western world has never had intellectual unity, and it is too much to expect, with several kinds of revolutions—scientific, technological, economic, social, political, psychological—taking place in our midst, that we can achieve intellectual unity in the twentieth century.

The Western world can be true to itself only by clearly recognizing the inherent diversity and pluralism of its heritage, a diversity and pluralism which have been increasing century after century. The diversities of European thought are more consistent with the Western heritage of freedom than is the growing uniformity of American thought. America's growing uniformity is essentially a betrayal of the American and of the Western heritage, and it is endangering the larger unity of the Western world.

&? Leninism and the Legacy of Western Imperialism

All major movements in history have their flows and ebbs, and by the 1960s the anti-imperialist revolutions were ebbing somewhat—their momenta were slowing down. There was less pressure to achieve industrialism in a hurry than there had been in the 1950s; there was more willingness to settle for a mechanization of agriculture and the extractive industries and for a rise in living standards less rapid than had been anticipated earlier. This article reflects the more chastened mood of the new nations of the underdeveloped areas.

Since this article was written, additional evidence has emerged to support the point of view that most of the new nations are still oriented largely to the West. When President Kwame Nkrumah of Ghana departed from an essentially Western orientation and moved to a neutralism slanted toward the Communist powers, he was ousted from office. When Kenya became independent in 1963, Jomo Kenyatta—the fiercest of revolutionaries and earlier the leader of the terroristic Mau Mau—became prime minister and later president. Yet Kenyatta in power has pursued a pro-Western policy. Returning American veterans of the wars in Vietnam and Laos frequently comment on the tenacity of the French culture in the countries carved out of old French Indo-China. Even the author, in the following article, written in 1962, underrated the continuing influence of the Belgians in the Republic of the Congo, for he indicated that the people of the old Belgian Congo would probably never feel kindly to the Belgians, whose colonial policy had been unenlightened and who supported the Katanga secession. Yet since the collapse of Katanga the leaders of the Republic have beaten a path to Brussels for credits, capital goods, administrators, and technicians.

Both leftists and rightists dislike intensely the view that Leninism has largely miscarried in the underdeveloped countries. Leftists resent any intimation that the old imperialism may have had a lasting, perhaps in some ways a beneficent, effect on the colonial peoples. Rightists are committed to a crisis mentality; they see Communist conspiracy everywhere; they sound the alarm that the new nations are teetering on the brink of Communism.

The Yale Review, LI (June 1962).

Among the leftists, Frantz Fanon and Jean-Paul Sartre are convinced that the "neo-colonial" peoples are seething with revolution and uniting in action against the West. They claim that the day of Europe and the European is finished. Their views are immensely popular among American black nationalists and the New Left generally. America's New Leftists see guerrillas bobbing up everywhere.

However, Herbert Marcuse, among the most scholarly of the New Leftists, in his *Essay on Liberation*, published in 1969, flatly contradicts Fanon and Sartre. Marcuse holds to the older Marxism and rejects the younger Leninism. He maintains that the chief thrust of Marxist revolution must come from the West, not from the neo-colonial peoples; that the "chain of exploitation must break at its strongest link."

There is something curiously nonhistorical about those historians who refuse to admit that European imperialism has left a permanent cultural impact on the former colonial peoples. All the other major imperialisms of history have made a lasting cultural imprint on the peoples they ruled. Why should the four centuries of European colonialism be the sole exception in this respect? The truth is that even if all the new nations carved out of the old European empires were to adopt Marxist systems, a residue of European culture would remain. Indeed, Marxism itself is a European invention.

■ ONE OF THE most curious paradoxes of our time is the way the West, while resisting the Communist powers, has itself been intellectually impressed by the Marxist interpretation of history, particularly twentieth-century history. Early in this century the high priests of Marxism made ambitious prophecies about what the forces of historical determinism had in store for us during the century, but we are now far enough along to see that history in practice is not working out according to their predictions. What is odd is the persistence on the part of non-Marxist Westerners in believing the Marxist myths.

When Communism triumphed in backward Russia and failed to come to power in industrially advanced Germany, this was contrary to Marxist prophecy. However, at the time of the Bolshevist Revolution, Lenin believed that a Communist world revolution was in the making, a revolution which would result in an international proletarian society. But this revolution miscarried after the First World War and again after the Second World War. The international proletarian society does not exist.

Communism has appeared only in a national form, and the various countries which have gone Communist have done so at different times and under different circumstances. In Russia and in China, Communism came largely as the result of indigenous national revolutions. In others, as in the Eastern European satellites, it has been imposed by Soviet imperialism.

Today, Communist countries differ from one another in history, conditions, cultures, stages of revolutionary development, and degrees of Marxist "orthodoxy" or "revisionism." Westerners, swayed by the Marxist myth, persist in exaggerating the solidarity of the Communist countries. Where there is "solidarity," as in the case of the Warsaw Pact countries, it is the solidarity of satellitism, a far cry from proletarian brotherhood.

Moreover, we in the West are still much impressed, consciously or unconsciously, by the Marxist doctrine of "capitalist imperialism." This emphasis on imperialism was largely Lenin's contribution to Marxism. Leninism holds that "dying capitalism" is kept alive by the economic exploitation of the colonial peoples, that these peoples will develop a great hatred for their masters and will eventually rise up against "the bourgeois imperialists," oust them, establish Communist societies of workers and peasants, and join the international proletarian society.

It is, of course, now apparent to all that the sweep of the anti-imperialist revolutions in Asia, the Middle East, and Africa is one of the most momentous and far-reaching facts of our time. It is also true that the Soviet system has marked attractions for peoples emerging from colonialism. These peoples do not have indigenous capital-supplying classes in sufficient numbers to finance industrial revolutions through private enterprise alone. They are impressed by Russia's industrial achievements, and they tend to regard the totalitarian economics of the Communists—curtailed consumption, forced savings, the regimentation of labor and collective capital into the heavy-goods industries—as an available, a sure, and a relatively self-reliant way to get an industrial revolution in a hurry.

However, most of the new nations are not adopting Communism. They dislike totalitarian dictatorship and the police state, they fear satellitism, and the advanced industrial countries are helping them achieve modernization and industrialization through less drastic methods. Instead, most of the new societies are moving in the direction of mixed economies, which combine privatism, collectivism, and welfarism in varying proportions.

Nevertheless, Westerners are still haunted by the prophecies of Lenin. This is reflected in the still rather general feeling that in the contest of East and West the Communists have a "natural" advantage among the underdeveloped peoples. Westerners too often exaggerate the appeal of totalitarian economics to the underdeveloped peoples and minimize the desire of these peoples for liberty and democracy. Many Westerners see Marxist influence in the government financing of some of the basic industrial developments in the new societies, although such financing arises from indigenous necessities which would exist even if there had never been a Communist revolution in Russia or China. In addition, out of a sense of guilt, Westerners fear that the colonial peoples, having built up hostility to the Western governing powers during the prerevolutionary and the revolutionary periods, will carry over that hostility into the postrevolutionary period, that the West for years to come must suffer the consequences of its "sins of imperialism."

The truth is the opposite. One of the most important advantages the West has in its contest with the Communist powers comes out of the fact that Western countries for a long period of time and until very recently ruled the colonial peoples of Asia and Africa. In spite of inherent contradictions, Western imperialism during its long sway planted Western concepts of liberty, democracy, and genuine education as distinguished from indoctrination. It modified Asian and African cultures in a Western direction. It left countless imprints of Western civilization. British or Dutch or French ways, more or less interwoven with indigenous ways, became the local and the familiar ways. The new societies in Asia and Africa cannot escape that impact, and most of their leaders do not wish to escape it. In India or Malaya or Egypt or Ghana or Nigeria, a Russian is a "foreigner" in a way an Englishman can never be, just as in Latin America the Russians and the Chinese are aliens in a sense that Spaniards—and even North Americans—are not.

Today, Communism is a greater threat in Latin America than it is in most of the new countries recently carved from the old Dutch, French, and British empires in Asia and Africa. Yet most of Latin America has not been under European rule for well over one hundred years. Latin Americans cannot blame their troubles on political imperialism, and unlike contemporary Africans and south Asians they cannot pin their hopes for remedial measures on national independence, for they have had that for generations. Therefore, the most impatient among them increasingly look to new economic and social systems for relief. However, the actual implementing of adequate foreign-aid programs in Latin America (long overdue) and the basic pull of Western culture there will probably prove decisive for the West in the battle with Communism.

It is highly significant that the one country (and a most important one it is) which seems to affirm the Leninist pattern—Red China—was never a political colony of a Western power. At a late stage of Western imperialism, large parts of China became economic spheres of influence of various European countries. These spheres were exploited economically, but imperialist governments were never established and the Western powers escaped political responsibilities. Their legal and political systems did not permeate the country. They did not build roads, schools, and clinics as they did in their political colonies. In not a single one of these economic spheres into which China was divided did any European culture penetrate in the way British culture penetrated India, Pakistan, Ceylon, Burma, and Malaya. (In Latin America, too, for the past century the advanced industrial countries have enjoyed immense economic advantages without political, social, and educational responsibilities.)

But in most of the new countries which have emerged from the old Dutch, French, and British empires—where imperialism was political, where it lasted long enough to penetrate culturally, where the governing power met its social and educational responsibilities to a reasonable degree, and where resistance to the national independence movements was not overprolonged or unusually violent—the Communist expectations are not being borne out.

The anti-imperialist revolutions have now been with us long enough to be put into something like true historical perspective. The process of a typical anti-imperialist revolution seems to run as follows.

First comes the long period of revolutionary agitation, with growing animosity toward the governing power. For the most part the revolutionary agitation is led by those who have been educated, not in Moscow, but in Britain, France, or some other Western country. Indeed, it is this Western education which has largely imbued the leaders with the idea of nationalism for their homeland.

The base of the national revolution gradually broadens. The conflict intensifies. At this stage many of the leaders pass through a more or less Marxist phase. They think of Marxism as a tool for winning the national struggle; they co-operate with out-and-out Communists; they flirt with Communism's totalitarian economics as an available method for the rapid industrialization of their country once independence is achieved.

Passive resistance, nonco-operation, sabotage, strikes, and violence increase. Armed conflict is imminent or actually arrives. Then the imperialist power bows to the inevitable, makes a settlement, and withdraws—sooner, with less resistance, and more gracefully than had been expected.

55

But much trade with the old governing country continues. Much Western capital remains. For more rapid industrial development, however, the government of the old imperial power extends grants and easy long-term credits to the new national government for technicians and capital developments.

Marxist sympathy and co-operation with the Communists ebb. The leaders of the new nation are now repelled by the drastic nature of totalitarian economics and the police state, and besides, with the economic help of the government of the old imperial power and other Western powers, particularly the United States, they see alternative and less painful ways to achieve fairly rapid industrialization. Moreover, the leaders have now become vested nationalists, nationalists in power, and they view Communism's theoretical internationalism and its actual satellitism as threats to their own position and their country's independence.

With the lapse of time the leaders and the intellectuals begin to put their revolution into perspective. They see that while the economic exploitation of colonialism undoubtedly played a part in their drive for independence, at bottom there was something broader and more basic—resentment of a people reduced to a subordinate place and at best second-class citizenship in their own homeland and a yearning for equal rights, individual respect, and human dignity.

When national independence has been won, the old animosity toward the former governing country subsides. The cultural impact of that country continues in myriad ways, some obvious and others more subtle. The attitudes of the indigenous people toward the people who once governed them soften, and with more time become mellow and even sentimental.

In short, after the revolution has been accepted, a reaction in favor of the old governing power usually sets in. Already this is quite marked in India, Burma, Ghana, and Nigeria with respect to Britain, and in Tunisia and most of Equatorial Africa with respect to France. The attitude of most people in the Philippines toward Americans is now downright affectionate.

However, it should be emphasized that this postrevolutionary reaction does not always take place, and under certain circumstances it should not be expected. It has not been strong in the countries carved from Old French Indo-China, for the French resistance to national independence movements in this area was needlessly prolonged and obstinate, and the Communists have been the gainers. (However, good will toward the French in Cambodia, Laos, and South Vietnam is not entirely lacking, and the peoples of these countries cannot escape French cultural influences.) A favorable Belgian reaction will not take place in the Congo, where imperialism was inhumanly brutal and unaccompanied by any sense of political, social, or educational responsibility. Nor will a favorable post-revolutionary reaction for Portugal ever develop among the long-oppressed peoples of Angola and Mozambique. But Algeria, in spite of the long and bloody struggle for independence there, will not be able to eliminate the cultural influences of more than a century of French rule, which penetrated deeply and often beneficently; and it is probable that within the next decade, with independence achieved, Algeria's good will toward France will equal that of Tunisia.[1]

A friendly reaction has taken place or is taking place among most of the peoples of the new nations which have emerged from the old French and British empires. It is particularly noticeable in most of the new African states. By the time the British and the French had to deal seriously with nationalist movements in Negro Africa they had learned the high art of liquidating empire. Timetables for self-government and then independence were deliberately worked out, and colonial administrators co-operated with nationalist leaders to these ends.

The independence of additional new states is now in the offing, and for some the dates for independence have already been designated. There are a few anticolonial revolutions yet to come. In Portugal's African colonies such revolutions are now gathering momentum. In Algeria such a revolution has long been under way and is now apparently reaching a successful climax.

However, the vast majority of the revolutions for national independence are now behind us. In most cases we are already in the postrevolutionary period—in the afterglow of Western imperialism—in which the cultural influences and the reservoir of good will left by the old imperialism will operate to the advantage of the West.

The changing attitudes involved in the various stages of the anti-imperialist revolutions may well be concretely illustrated in the backgrounds

1. Since independence, Algeria has maintained in world politics a neutralism friendly to the left, but certain cultural ties with France persist. As events have turned out, the author underestimated the vitality of his own thesis, for he exaggerated the post-Katanga hostility of the Congolese to the Belgians; the Republic of the Congo is still dependent on the Belgians in many ways. In the light of this Congolese-Belgian example, it may be that even the peoples of Angola and Mozambique will work out cordial relations with Portugal once they achieve their independence.

and political careers of the leaders of those revolutions. What is true of the outstanding leaders is in the main true of the numerous less conspicuous leaders, those known at home but not abroad.

Back in the 1920s, when the Indian Revolution was intensifying, Nehru was much attracted to Leninism, but later he returned to his earlier Fabianism. Nehru spent many years as a student in Britain, and today many of his basic cultural attitudes and certainly his twin political ideals of liberalism and socialism reflect British influences. U Nu, who led Burma's fight for independence and is his nation's hero whether in or out of power, never went much beyond socialism. Today his views are a blend of Fabianism and devout Buddhism, and he makes no secret of his warm regard for Britain and his fraternal friendship for the British Labor Party.

Most leaders in Middle Eastern countries have been educated in Britain or France, or served in the British or French army, or had army training that was Western oriented. Pakistan's current leader, Mohammed Ayub Khan, is a product of the Royal Military College at Sandhurst. For many years he was an officer in the British army. Observers often say that he looks and acts more British than the British—"mustache, swagger stick, stiff upper lip and all that." Abdul Karim Kassem of Iraq, who led the movement to overthrow King Faisal and take his country out of the pro-Western Baghdad Pact alliance, is a product of British army training, and he pursued the senior officers' course in England. He had not proved as anti-Western as many feared he would. Even Nasser came up under British-oriented army training.

Besides, Nasser is a seriously religious man and a dedicated Moslem. Moslem populations have been tenaciously resistant to Communism, not only in Arab countries but also in Turkey, Iran, Pakistan, Burma, Thailand, Malaya, and Indonesia. Much of the West's trouble with the Arab world has come from Arab resentment over the West's sponsoring of Israel, and in time, in spite of Arab pronouncements to the contrary, this resentment will soften, provided the Israeli population does not grow beyond Israel's territorial capacity to sustain it.

Habib Bourguiba, the President of Tunisia, was educated at the University of Paris, married a French girl, and has never hidden his admiration for French culture. He has always insisted that his differences with the French were primarily political. In addition to leading the fight for the independence of his own country, Bourguiba supported Morocco's claims to independence, and he has backed the Algerian revolution. In spite of all this and his continuing difficulties with the French, Bourguiba leaves no doubt of his sympathies for the West and his affection for French civilization.

Felix Houphouet-Boigny, now President of the Republic of the Ivory Coast, went through a Marxist phase, and in the late 1940s his revolutionary party co-operated closely with the Communists. However, Houphouet later broke with the Communists, and since the independence of his country he has grown more moderate. He has worked in close harmony with the French, and he has been an ardent supporter of the French Community, which resembles the British Commonwealth. He encourages former French colonies to remain in the Community and continues to urge his close friend, President Sékou Touré of Guinea, to bring Guinea closer to France. (Guinea is the only former French colony in Negro Africa to break all official French ties.) Touré rose to revolutionary leadership in his country because of his effective work in the trade-union movement and his capacity to organize labor for political purposes. There are now definite signs that Touré's long flirtation with Marxism is coming to an end. In all probability, Guinea will not be able to escape its basic French orientation any more than its fellow French-speaking neighbors.

President Kwame Nkrumah of Ghana went to college in the United States and Britain and in the 1940s, when his country's independence movement was getting well under way, seems to have flirted with Marxism. Later he became more conservative, worked closely with the British to obtain his country's independence in a peaceful way, and since independence has co-operated with the British within the Commonwealth.[2] Tom Mboya, leader of the independence movement in Kenya, is the product of the fierce racial strife of that colony. He has been a tough trades-union organizer and proclaims "Africa for the Africans." But even Mboya interrupted his revolutionary career in 1956 to study in England, and today, with Kenya's independence in sight, he shows signs of becoming more conciliatory. The two most ardent revolutionaries in Africa today—Kenneth Kaunda of Northern Rhodesia and Joshua Nkomo of Southern Rhodesia—are followers of Gandhi and not Marx. Prime Minister Balewa of large and populous Nigeria, President Tubman of Liberia, and Julius Nyerere of Tanganyika (who holds a degree from the University of Edinburgh) are all stoutly pro-Western and democratic gradualists in philosophy and practice.

Thus it is apparent that many—though not all—of the leaders of the

2. When Nkrumah moved away from a British-Western orientation and toward a neutralism slanted to the Communist powers, he was toppled by his own people.

new nations have passed through moderately Marxist phases. With national independence achieved or within sight, most of these have dropped their Marxism. Almost without exception the leaders have had Western educations, as have the classes from which the leaders have mostly been drawn. Even the local and primary schools in the old colonies have radiated Western culture. And even those large segments of the population which have had little or no formal education know something of the language of the old governing power, are accustomed to dealing in terms of that country's monetary unit, are used to the make and style of goods from that country, and continue in considerable measure to buy from and sell to that country. Political institutions and ideals have been greatly influenced by the old governing power. These are only the more obvious manifestations of the continuing impact of the old imperialism on the new countries, for that impact asserts itself in innumerable and subtle ways.

The old imperialism continues to manifest itself in even the political combinations the new countries make. All but one of France's former colonies in Negro Africa are either members of the French Community or closely associated with France through bilateral treaties. Most of Britain's former colonies are members of the Commonwealth. After all the discussion of a West African federation, it seems likely that if an effective federation in that area is achieved it will take the forms of a West African federation of French-oriented nations and a West African federation of British-oriented nations. Among the difficulties of the abortive United Arab Republic was the fact that for many years Egypt was in the British colonial orbit and Syria in the French.

Most of the economic collectivism in the underdeveloped nations—that is, government initiating and financing of enterprises basic to industrialization—is not Marxist. It is indigenous, and it grows out of pragmatic necessity. Where there is a social philosophy behind it, as in India and Burma, it approximates Fabian socialism, borrowed from British thinkers.

Had the Communists or the thoroughgoing pre-World War I Socialists come to power in European countries with important overseas colonies, there is little doubt that Marxism would have had a most powerful impact on the colonies and would have influenced them in the building of their institutions after independence. The miscarriage of Marxist revolution in Western Europe has had decisive consequences in the former colonial areas.

In this sense the older Marxism was much more realistic than the younger Leninism. Marxism expected a thoroughgoing Socialism or Communism to triumph first in the advanced industrial countries of Western Europe and then from Europe to influence the colonial areas. Leninism gave rise to the belief that the Communists could win the West through a prior victory in the colonial areas. We now see that the miscarriage of Marxism in Western Europe resulted in leaving the Marxists with insufficient prestige and too few Western cultural carriers to win in the colonial areas.

Now, there is no doubt that the peoples emerging from colonialism resent the continuation of colonialism in such places as Algeria, Angola, and Mozambique. They are also irritated by vestiges of imperialism, whether these are foreign air bases, oil and other special business concessions to foreign nationals, formal economic preferences for the old imperial power, or enclaves or nearby territory they regard as still nationally unredeemed. They also bristle at the still all-too-frequent assumption that their foreign policies are for sale or are susceptible of "management" by the West. These peoples, too, have an admiration for the industrial and technological achievements of Soviet Russia, and they are willing, often eager, to take economic aid from both the West and the Communist powers.

Nor can the West afford to take the friendship of the new nations for granted. They want socially effective technical and economic aid for their industrial developments. (Western economic aid which is administered by oligarchic national governments and which fails to result in an increase in wealth-producing enterprises and a rise in mass living standards is worse than useless; it backfires and causes disillusionment.) They want understanding and respect. They are sensitive to racism and quickly detect inconsistencies between the West's democratic professions and its practices. They abhor the intensification of the power conflict, the nuclear arms race, and the mischief done to outsiders by atomic testing.

It is true that most of the new nations are neutralist—refuse to line up with either of the opposing power blocs. Some, organized into a distinct bloc of their own, assert a positive neutralism. Others are opposed to all permanent and formal blocs, and among the new African nations only Ghana has aligned definitely with the Nehru-Sukarno-Nasser-Tito combination.

Because most of the new nations are neutralist in international politics we tend to overlook the fact that they are preponderantly Western in their orientation. They are not Communist. They are not police states. Even those which seriously abridge democratic rights and civil liberties cannot accurately be termed totalitarian. Their economic systems are not mono-lithic but mixed and fluid. Their intercourse is far more with the West than with Communist countries. The outside cultural influences continue to be far more Western than they are Russian or Chinese.

Now, of course, it is quite possible that some of the new nations now non-Communist may eventually go Communist. Whether this does or does not happen will depend largely on the adequacy and wisdom of Western policies with respect to the underdeveloped peoples. At this juncture in their history most of these peoples are putting their faith in their new governments and their new nationalism. Should these fail them, some might seek more drastic social and economic changes and turn to Communism. There is the likelihood that additional new nations, failing to fulfill national expectations in a hurry, will fall into the hands of transparently nonpopular, antidemocratic oligarchic-military governments, and that this in turn will increase the chances of ultimate Communist take-overs in such nations.

However, all of this departs widely from the way Lenin envisaged the historical process. We now know that after their anti-imperialist revolutions most of the old colonial peoples did not turn their backs on the old imperialist powers and that they did not join an international proletarian society. Instead, their revolutions became national revolutions and after independence had been won the former colonial peoples looked to the West for help and stability. They gave the West another chance.

Of this we may be sure: all of the new non-Communist countries will not go Communist, and if in the future some should go Communist, their new revolutions, their Communist revolutions, will take national forms and be conditioned by their national cultures. And those national cultures, in turn, have been durably affected by Western civilization during the long period of imperialism.

In most of the new nations, nationalist parties, largely peasant in mass composition, have come or are coming or will in the future come to power. These parties sometimes use Marxist terminology to justify indigenous collectivist necessities, but in practice what they do or propose to do is so far removed from a genuine Marxism that Communists do not recognize it as Marxist at all. These parties invariably emphasize a revival of indigenous language and culture, and from now on Western cultural influences will be muted and modified.

However, these Western influences cannot be eradicated; the nationalist parties tacitly accept much of the West's cultural impact; indeed, the passionate drive of the underdeveloped peoples to nationalism, industrialism, and rational living standards dramatically attests the continuing vitality of Western values. The Communists, of course, accept and exaggerate Western industrial and rational values; what they reject are other aspects of the Western heritage such as democracy, civil liberties, and intellectual freedom. But the old colonial areas have been exposed to many aspects of Western culture, including the nonmaterial, and the nonmaterial aspects are also having a continuing effect. Gandhi's life exemplifies a blend of the deep spiritual experiences of both the East and the West, and Nehru has been profoundly influenced by the West's humane values as well as its rational ones. The record up to now shows that most underdeveloped peoples clearly prefer to achieve industrialization and modernization in a nontotalitarian and humane way and that only if this way demonstrably fails will they resort to drastic Communist economies.

For the underdeveloped peoples, their leaders, and their nationalist mass parties, there is magic in the Western term democracy, but the achieving of effective democracy must necessarily be difficult. Democracy will suffer many ups and downs along the way. It is easy to sneer at Nasser's "presidential democracy," at Ayub Khan's "basic democracy," and at Sukarno's "guided democracy," but such slogans more often than not exemplify honest and realistic attempts to move in the direction of democracy. Even India's Congress Party and its government must be highly paternalistic at this stage of India's journey to democracy.

Western imperialism, then, left among the old colonial peoples both material and nonmaterial Western values, and these, blended with a new emphasis on indigenous languages and folkways, constitute the chief dynamics of their contemporary history.

It is indeed strange that so many Westerners still take for granted that the Communists have the current advantage over the West among peoples recently emerged from colonialism, still believe essentially in the "realism" of the Leninist doctrine of imperialism, with its monistic economic interpretation of the "inevitable" trend of history in the old colonial areas and its prediction that the former colonial peoples would carry over into the postrevolutionary world an implacable hatred of the "bourgeois" civilization of the West. One explanation for the persistence of this myth is that many Americans think that a mixed economy, which combines indigenous collectivism, state enterprise, private enterprise, and state welfarism, is the same thing as Communism; they see the spread of mixed economies as the spread of Communism, whereas of course these fluid mixed economies differ sharply from totalitarian Communist economies. The actual course of the anti-imperialist revolutions since 1945 strongly indicates that it is the West which has the advantage, and the most basic reason for that advantage is the continuing vitality of Western culture and the permanent cultural impact left by Western imperialism itself in the old colonial areas.

Throughout history, long political rule has almost always resulted in lasting cultural influences. These influences have generally been more obvious when the colonies or provinces ruled were largely composed of people transplanted from the governing country itself, but sheer conquest of a people of one culture by people of another, when the period of imperial rule covers considerable time, produces significant and enduring cultural changes.

The anti-imperialist revolutions of the late eighteenth and early nineteenth centuries—the Anglo-American and the Latin-American revolutions—severed the political ties of the American colonies from their European rulers, but cultural affinities remained. In spite of the War of 1812 and the propensity of Americans during the nineteenth century to "twist the lion's tail," American-British relations in economic and cultural matters remained close, and after more than a century of political separation, common cultural ties were among the most important forces bringing the United States and Britain together in twentieth-century world politics. Even among the Indians and mestizos of Latin America today, the Spanish culture largely prevails, and most of the educated among them find their spiritual homes in Rome and Madrid and in Spanish humanism.

The tenacity of the cultural impact left by the historical imperialisms is impressive. Even at the beginning of this century the influences of the Greek colonies of antiquity in Asia Minor and the Black Sea area were discernible. Stendhal, in his delightful accounts of his travels in Italy, noted as late as the early nineteenth century that in the southern part of the peninsula—in old Magna Graecia—Italian life and attitudes still revealed marked Greek influences. Even today some of the significant characteristics which distinguish European peoples from one another have their origins in the extent of the Roman conquests and the length of time the various areas were under Roman occupation.

European imperialism, which began with the Age of Discovery and Exploration in the late fifteenth century, is now drawing to its close. Unlike the other major imperialisms of history, it has in the main been liquidated with a minimum of violence and bloodshed. But like the other major imperialisms, its cultural influences will be felt through the centuries.

Today, in its struggle with the Communist powers, the West, notwithstanding a stereotype to the contrary, is the beneficiary of cultural linkages it forged with the colonial peoples during its long years of imperial rule.

～ Our Post-Crisis World

The following article argues that the earth-shaking crises of this century are behind us and not ahead of us; that the nuclear revolution in warfare is making the powers skittish about all wars, little as well as big; that this fear of war is freezing the international status quo; and that in effect the Americans have won the Cold War but do not yet realize it.

Nearly seven years after this article was written, James Reston wrote: "Never in human history have two sovereign nations, with such contradictory philosophies of human life and such apocalyptic military power, faced one another over so large an area of the earth and even the universe, as the United States and the Soviet Union in the last generation. Never have such hostile powers with such conflicting interests faced so many provocations to let events get out of hand in Berlin, Hungary, Korea, the Middle East, Cuba, and Vietnam, and go on to the big war. But always, in this phase of Washington-Moscow control of world events, one or the other of the two major governments has pulled back from major war at the point of crisis."

Today's international relations are paradoxical. In a number of ways, as the following article explains, the world has been moving to a larger pluralism. At the same time, the two nuclear superpowers, the United States and the Soviet Union, continue to predominate, and the trend is to freeze the status quo.

The mutual nuclear deterrents compel the two superpowers to coexist in an uneasy *modus vivendi*. These deterrents force them to put restraints on themselves, so that while they sometimes engage in brinkmanship they always recoil from war.

Both superpowers are keenly aware that the chief cause of the big-power wars of the past has been a threat to an existing balance of power. They are also keenly aware that any big-power war today means nuclear war. Therefore, they are extremely touchy about any change in the current balance between them, any change in the status quo that might affect that balance. Hence they intervene in "little military occupations" or "little wars" to maintain the status quo—as the Soviets have done in Hungary and in Czechoslovakia, and as the Americans have done in the Dominican Republic, in South Korea, and

The American Scholar, XXXIII (Winter 1963–1964).

in South Vietnam. In the absence of other kinds of stabilization, even the difficulties of the Americans in the Vietnam war are not likely to dissuade them from further military interventions to preserve the status quo, although in the future they are likely to depend less on American draftees and more on indigenous armies supplied with American equipment and on American volunteer soldiers.

Since action to change the status quo is much more risky than action to preserve it, both superpowers desist from direct intervention to change the status quo. Hence the United States held aloof from helping rebellious elements in East Germany, Poland, Hungary, and Czechoslovakia, and it has abstained from attempting to topple the Communist regimes in North Korea and North Vietnam, while the Russians have steered clear of any direct involvement in the "war of liberation" in South Vietnam.

Has revolution been foreclosed in large portions of the world? With each of the superpowers squelching significant departures from the status quo and backing away from giving direct help to the forces of change, how can revolutions succeed? In the past, revolutions have usually won out either because of outside help, as in the case of French aid to the American Revolution, or because of a general collapse during or following a big international war, as in the cases of the Russian Revolution and the Chinese Communist Revolution; but big international wars are precisely the kind of wars the nuclear deterrents are banishing.

The superpowers are not only putting restrictions on themselves and restraints on each other, but they sometimes co-operate—usually tacitly but occasionally overtly—to put pressures on other countries, with a view to stabilizing situations and preventing crises that might affect themselves. There was such co-operation to liquidate the Suez crisis of 1956, to check China in its foray into the Himalayan borderlands of India, to restrain India and Pakistan in their Kashmir dispute, to neutralize Laos, and to scuttle the firebrand Sukarno. The continuing crisis in the Middle East may persuade the United States and the Soviet Union to co-operate again in imposing peace there.

It may well be that the two superpowers, increasingly aware of the burdens and perils of the "little" unilateral military interventions to maintain the status quo, will move eventually to co-operate in establishing neutralization arrangements covering certain of the tension-provoking areas, notably old Indo-China and the Middle East. The two superpowers may come to regard this concerted form of imposed immobility as more advantageous to themselves, in some areas, than the present-day unilateral interventions.

If the superpowers were not restrained by the nuclear deterrents, they probably would engage in conventional war on a large scale. Then all the pent-up forces of change would have their opportunities to take advantage of such a war to explode the status quo—the Latin Americans and the peoples of Southeast Asia to have their social revolutions, the democratic Socialists in the Soviet's Eastern European satellites to oust the Communists, and Vietnamese, Korean, and German nationalists to reunify their countries. Not all of these movements would win; there would be divisive elements in all these situations; but at least the forces of change would have their chances.

However, the nuclear revolution in warfare cannot be undone; those forces throttled by it must work in more localized and circumspect ways than were used in the wars and revolutions of the past; but there is cause for optimism that the scourge of big war has almost certainly been eliminated and that the same technological revolution which produced the *pax atomica* may also work to better the conditions of life everywhere and thus make the pressures for revolutions less compelling than they were in the past.

We must beware of claiming "inevitability" for any historical trend, of putting history in a strait-jacket. Before World War I, Sir Norman Angell predicted that there would never again be another big international war, that industrialism and modern technology had made war too destructive and horrible for human endurance. And after that came World War I and then World War II. Nevertheless, before this century is out, Sir Norman may have become one of the world's true prophets. He correctly observed the *trend*; but he miscalculated the *degree* of revolutionary technology, the *extent* of "total war," required to chasten man's penchant for war.

Until man is further chastened and international peace machinery institutionalized, it would be well to limit nuclear weapons (to reduce the intolerable burden of their present-scale maintenance) and by all means to keep them in balance as between the two superpowers, but not to abolish them; for it is the mutual nuclear deterrents which are working to stabilize the world, albeit at the awful cost of a rigidified status quo.

■ IN ALL PROBABILITY the peak crises of our century are behind us. In long historical perspective it increasingly appears that we of the last half of the twentieth century are living, not in the crisis world, but in the post-crisis world of reconstruction and stabilization.

This view will be greeted with skepticism. If it is true, future generations will regard the Kennedy inaugural address as curiously off-key and not the Lincoln repeat that many Americans saw in it at the time.

The first part of our century was one of repeated high crises and sweeping change—total world wars, basic revolutions, devastating inflations and deflations, the momentous atomic breakthrough. The latter part of the century promises to be one of stabilization, of constructive adjustment to the cataclysmic events of the first half.

But we carry with us the crisis mentality developed before 1950; indeed, we have intensified it. And obviously we *are* still beset with dangers. The Soviet Union is a tremendous military power, and China will soon become one. Americans appear to be trapped in a fear psychosis and saddled with a potentially dangerous military-industrial power complex. Some of the anticolonial revolutions may yet pass through a more radical second phase and be captured by Communism. The threat of thermonuclear war, the balance of terror, hangs over our heads.

Certain basic trends, however, are becoming increasingly clear: that Communism has been rather effectively checked and that the forces mobilized (largely by the United States) to keep it in check are stronger and of greater military and nonmilitary variety than ever before; that Communist societies themselves are becoming less aggressively revolutionary as they grow older and industrialize; that the anti-imperialist revolutions are about to pass into history without vindicating Lenin's prediction that they would merge with Communism; that Communism itself is not the unified movement it was prophesied and feared it would be; that the world is not going to become a Communist or any other kind of monolith; that a pluralism of many new societies with pragmatically fluid economies is emerging; that today the modified capitalist systems in the older industrial countries are operating much more smoothly than the "purer" capitalism of the early years of the century; that the nuclear balance of terror is in fact producing a *pax atomica* which, among other causes, almost certainly precludes the climactic Communist-Capitalist Apocalypse of Marxist dream.

The central political and social challenges of our century have been Marxism in its various forms and Nazi-Fascism, an extremist counter-movement to Marxism.

On the eve of World War I, despite the internal ideological differences within the Marxist parties of the Second International, these parties in Western Europe had popular strength, revolutionary élan, and confidence that the future belonged to them. The failure of thoroughgoing Marxist revolutions to materialize in Western Europe during and immediately after World War I is one of the decisive facts of our century, perhaps the most decisive.

In 1914, at the outbreak of the war, European Marxists rudely discovered that their nationalism was stronger than their Socialism, and most of them supported the national war efforts. Outside Russia, Marxists did not take advantage of the confusion and hardships of the first total war in history to convert "imperialist" war into class war. Never again were they to have so favorable an opportunity.

Encouraged from Moscow and exploiting the misery and dazed condition of the populations immediately following the war, Communism was indeed a formidable threat in 1919. Several times that year the Communists attempted to seize Berlin and overthrow the new postwar German government. For several months Communist dictatorships held power in Bavaria and in Hungary. In the Italian elections that year, the revolutionary Marxist Socialists (not yet split into Communists and Socialists as they would be after the Livorno Congress of 1921) came near to taking over the government. But in the end all revolutionary Marxist attempts in Western Europe proved abortive.

At various times during the 1920s and the 1930s, democratic Socialist parties held power, more or less, in Germany, Austria, Britain, and France; but nowhere did they tackle power with the revolutionary enthusiasm of pre-1914 days, and everywhere the reforms they enacted fell far short of establishing socialist societies.

Without doubt the Marxists always underestimated the tenacious strength of capitalism itself. But other factors also contributed to the failure of the Marxists: their sense of humiliation and frustration at having missed their opportunities during World War I, their inability to recapture their prewar revolutionary élan, and the bitter split between the democratic Socialists of the Second International and the Communists of the Third International. The methods of the Bolshevist Revolution chilled the democratic Socialists of Western Europe and dealt West European Marxism a fatal blow. Thereafter Socialists and Communists fought each other as much as they fought the non-Marxists. And meantime the capitalist economies of most of the advanced industrial countries were being so modified by Keynesianism, the welfare state, and a widening of their public sectors as to blunt their traditional inequities and consequently the opportunities for drastic social revolutions.

The next great crisis of our turbulent century was the Nazi expansion which came to a head in the years of World War II. There is much more complete agreement on this being a historical crisis of the first magnitude than on the earlier crisis that involved "merely" the miscarriage of Marxism in Western Europe. These are those, Arnold Tonybee among them, who see in the conflict of capitalist democracy and Marxist Communism only a struggle between two aspects of Western civilization, like the conflict between Protestantism and Catholicism in the sixteenth and seventeenth centuries, fierce and bitter in its day but in the long run not involving the survival of civilization or even of Western civilization. However, capitalists and democrats, Socialists and Communists, Churchillian imperialists and Nehru anti-imperialists, the world's religions, and those who view the contest of capitalist democracy and Marxism with considerable detachment—all generally agree that the war against Nazism represented one of civilization against nihilism.

The third great crisis of our era was the threat of Communist takeover in Western Europe following World War II. So rapidly have events moved since 1945 that we are likely to forget the revolutionary mood of the immediate postwar years. Europe's middle classes, battered by the earlier world war, then by the inflation of the 1920s, then by the depression of the 1930s, and then by the Second World War, seemed in the mood for revolutionary change. There was a bitter feeling that powerful vested interests had been collaborators and "gravediggers" and had sold out their own peoples to the Hitlerians in the hope of "escaping Communism." Communists had taken an active and often the most aggressive leadership in the various national underground movements. Even those who detested Marxism proclaimed a Christian socialism to defeat it. Even De Gasperi, leader of the opposition to Marxism in Italy, was reported in the first days of the liberation to have asked publicly: "Was not Christ the first communist?"

But again, Communist and even thoroughgoing Socialist revolutions miscarried. In Western Europe, in the years between 1945 and 1950, there were agrarian reforms and some nationalization in Italy, considerable nationalization in France, and a "Socialist Revolution" in Britain. But no country in Western Europe achieved a socialist revolution in the sense the pre-1914 Marxists envisaged it; no Western society became a socialist society. And in West Germany there was even some movement away from the nationalization and socialization that had been achieved during the Weimar Republic and a trend in the direction of a free enterprise society in the American sense. Over-all, what emerged was modified capitalism, not socialism.

We have come to regard the impact on Western Europe of American capitalism, particularly as this manifested itself in the Marshall Plan, as the most decisive factor in preventing the spread of Communism or the triumph of some brand of thoroughgoing Socialism in Western Europe. This is important and never to be discounted, but it is possible that we overrated the prospects of Marxist victory and underrated the strength of European capitalism itself. The hold of the past was tenacious, and even during the years between the two world wars—yes, even during the depression years—Europeans had some taste of what modern technology, reformism, state welfarism, and Keynesianism, without drastic revolution, might accomplish. The old split between the Communists and Socialists had not healed, and indeed the Stalinist terror and purges of the 1930s and the Soviet expansion in Eastern Europe following World War II frightened not only the non-Marxists but also the Socialists.

Even in the years immediately following World War II, the Communists in France were able to win only about one fourth of the French electorate, and many who voted Communist were not Marxists but merely casting protest votes. In Italy, the one country where during the late 1940s and the 1950s the Communists and majority Socialists were able to maintain an uneasy electoral alliance, the combined leftist vote did not rise above one third of the electorate, and Italian Marxists failed to regain the enthusiasm and drive of their 1919–1920 crest. In industrial West Germany the Communist Party did not revive at all, and the Social Democrats had lost the old momentum of pre-Hitler days.

And this leads to a conclusion scarcely ever articulated and frankly distasteful to liberals, who believe that ideas cannot be stopped by force: that the important and permanent historical impact of the Fascist and Nazi movements lies in their over-all weakening of Marxist parties in Western Europe. Repression and persecution *were* effective. Marxist parties in countries governed by rightist totalitarian dictatorships were either seriously enervated (as in Italy) or decisively impaired (as in Germany and Spain) by the persecutions carried on by right-wing despotisms.

Thus the brutal liquidating of Communist, Socialist, and labor leaders by rightist dictators, together with the positive and beneficent impact of American capitalism following World War II, created a situation that led to the miscarriage of Marxism's second great challenge in Western Europe.

As we move further into this century and get a larger and truer perspective of it, it becomes clearer that the failure in Western Europe of genuine Marxist revolution of either the Communist or Socialist variety in the years from 1914 to 1920 was pivotal; after those years what opportunity there had been for Marxist triumph never again appeared with equal strength. Lenin, in the very marrow of his bones, seems to have sensed that this would turn out to be the case. After the triumph of the Bolshevist Revolution, Lenin, of course, never gave this public expression, but increasingly he put the emphasis on the importance of winning the colonial areas to Communism.

Leninism, as distinguished from the older Marxism, encouraged the notion that even if Communist victory were delayed in the industrial countries, it would win in the colonial areas, thus assuring ultimate triumph in the West. Leninism emphasized that the colonial peoples would expel their imperialist masters in violent revolutions, establish Communism, join an international proletarian society, and carry into the postimperialist world their hatred of their imperialist exploiters.

Since World War II, anti-imperialist revolutions have indeed swept Asia, the Middle East, and Africa, and scores and scores of new nations have been born. The resistance of the imperialist powers was much less tenacious than had been predicted. No other major imperialism in history was ever liquidated in so orderly a fashion. In many cases the British and the French practiced "a creative abdication," prepared their colonial peoples for self-government, and retreated in phased stages according to timetables agreed to by the leaders of the nationalist movements.

By now most of the anti-imperialist revolutions have already passed into history. Few European colonies remain to be nationally redeemed. And one stupendous fact stands out: the anti-imperialist revolutions did not merge with Communism. The only European political colony that in the process of its nationalist revolution went Communist was little North Vietnam. This is a sobering and humiliating fact to Communists, for if they can take no better advantage of "revolutionary situations" in a tumultuously revolutionary age, how can they expect to win when the revolutionary surge subsides?

Communism has indeed made conquests in some of the economically underdeveloped countries—in North Korea, Mongolia, and, above all, China. None of these was a European colony. At a relatively late stage of Western imperialism, China was carved by the European powers into economic spheres of influence and there was much exploitation. The Western powers enjoyed enormous economic privileges without assuming governmental and social responsibilities. They did not build roads and schools or adopt public health measures as they did in their political colonies. In no European sphere in China did a European culture penetrate to the degree, for instance, that British culture penetrated India.

During the coming decade a most likely area of social revolutionary penetration is Latin America, which for the most part has never broken out of the feudal land concentration and social caste system inherited from Spanish and Portuguese colonial times. Today the impoverished and underprivileged masses are pervasively challenging the old order as never before. Unlike the peoples of Asia, the Middle East, and Africa, Latin Americans can pin no great hopes for betterment on nationalism and political independence, for they achieved these a century and a half ago and have been repeatedly betrayed by their own governments, leaders, and ruling classes. Therefore they may turn to more drastic revolutionary movements.

Even in Latin America, however, the odds are against Communism. Despite the population explosion, Latin American countries still do not have massive populations; their man-land ratio is still low. They already have a wider industrial base than most of the other underdeveloped societies. During recent decades they have been making some industrial and social progress. Paradoxically, Castro's career may result in moderating Latin American revolutions. The Castro challenge alerted the United States to the necessity of the Alliance for Progress. Also, by announcing himself a Communist, Castro threw away the magic symbolism of leader of indigenous Latin American revolution and checked the growth of his own following throughout Latin America.

There is no doubt that Latin America is in for trouble. The ruling oligarchies in Latin American countries will bitterly fight change. There will be more demagogues of the Perón stripe, who, promising the good things of advanced industrialism without the savings and sacrifices that alone make those things possible (a mistake both orthodox capitalists and Communists avoid), thereby precipitate headlong inflation, bankruptcy, and the blighting of mass expectations. The Communists will foment and exploit the discontent. But here and there reforms will be made by the democratic middle way, aided by the Alliance for Progress, and these reforms will accumulate. The mushrooming of Chile's Christian Democratic party, which is to the left of the old liberal parties and in the tradition of Catholic social reform, may be a portent of the future. Where sweeping social revolution takes place it is likely to be in the tradition of the Mexican Revolution or somewhat to the left of the *apristas* in Peru. The Communists would have a part in such revolutions—sometimes as constructive helpers, sometimes as nuisances, and sometimes as saboteurs—but the successful revolutions are likely to be of the home-grown variety, much like the Nasser social revolutions in the Arab countries, and not Communist ones.

Moreover, the Americans would risk all-out nuclear war to prevent a massive Communist penetration of the Western Hemisphere, a fact the Russians well know—an illustration of the developing tacit understanding of East and West to permit no wide departure from the international status quo. (By 1963 it appeared that, while in the Communist bloc there were "many roads to socialism" and in the non-Communist bloc there could be social revolutions if these were not Communist, the non-Communist countries would not actively intervene to push Communist countries to non-Communism, the Communist countries would not directly intervene to push non-Communist countries to Communism, and both East and West would make agreements to keep neutralist countries neutralist, as in Laos.)

It was in the colonies and protectorates of the European powers that hatred of foreign rule, economic exploitation, and racial arrogance was expected to make the Communist thrust most sweeping. Despite colonial resentments, however, and the powerful pragmatic attraction of Communism as a way of achieving an industrial revolution in a hurry, the spread of Communism to the former European colonies has been negligible.

The former colonies have not even cut their ties to the old imperial powers to the degree usually supposed. For the most part, the new nations voluntarily maintain much trade and intercourse and considerable fiscal connection with their former rulers. Even all the political ties have not been severed. Former British colonies are members of the Commonwealth, and Britain is associated with the Colombo countries of southern Asia for purposes of their economic development. Some of France's colonies in Africa are members of the French Community, and most of the others are still intimately connected with France in various ways by bilateral treaties. The twelve so-called Brazzaville countries are associated, through France, with the European Common Market.

The Western cultural impact on the populations at large, the result of long years of imperial rule, manifests itself in many overt and subtle ways. This cultural impact is even stronger on the educated classes. Most of the rulers of the new countries have had Western educations, and today many more of the rising generation of leaders and technicians are getting their educations in the West than in the Communist countries. The nationalism learned from the West makes the new nations sensitive to Com-

75

munist satellitism. The democracy and civil liberties learned from the West make them resistant to totalitarian police states. Even when new nations adopt some form of authoritarianism, this falls far short of the omnivorous controls of a Communist society. Most of the new nations are attempting to get rapid industrialization by methods that combine collectivism and privatism, government initiative and financing in the large basic projects and private enterprise in the myriad of smaller ones. The new nations find that generally the economic aid, the capital goods, and the technicians furnished them by the West are superior in both quantity and quality to those furnished them by the Communist countries.

In short, in the old colonial areas of the world, the influence of the West is still much greater than the influence of the Communist countries. Americans often overlook this basic fact. Because the new nations have not adopted free enterprise economic systems in the American sense but are developing mixed economies which combine both state enterprise and private enterprise, many Americans are inclined to think of the new nations as "Communist." Because most of the new nations have adopted neutralist positions in world politics and refuse to join America's system of military alliances, many Americans suspect that the new nations are "pro-Communist." But the truth is that the new nations have escaped Communism; they are often not anti-Communist but they are certainly non-Communist; and they enjoy wider and closer relations with the non-Communist countries than with the Communist ones.

It doubtless would have been another story if Marxist revolution had triumphed in Western Europe. Had Communism or even the thorough-going Socialism envisaged by the pre World War I Socialist leaders taken over important Western societies, the impact of this on the European colonies would have been pervasive and profound. In fact, Marxist influence on the nationalist parties fighting in the colonies for independence usually rose and fell with the fortunes and prospects of the Marxist parties in Europe.

For instance, during the years from 1945 to 1948, when the influence of the Communist Party in France was at its postwar peak and when for a time Communists were members of French cabinets, the nationalist parties in France's African colonies veered closer to Marxism. Even so moderate a nationalist leader as Félix Houphouet-Boigny in the Ivory Coast increasingly used proletarian appeals in recruiting membership, and he himself co-operated closely with Communist leaders in Paris. But when the domestic influence of the French Communists declined, so did the influence of Marxism on Houphouet-Boigny's party and other anti-imperialist parties in France's Negro colonies in Africa.

As the years roll by and we get a longer perspective on the twentieth century, the miscarriage of Marxism in Western Europe after World War I and again after World War II looms as truly decisive for world history. Failing to win Western Europe, Marxism was left with insufficient prestige and too few Western political and cultural carriers to win the colonial areas. In this sense, the earlier Marxism, which envisaged the triumph of Communist or Socialist revolution first in the advanced industrial countries of Europe and after that in the colonial areas, was far more realistic than the later Leninism.

All of this is not to say that the old colonial areas have no present-day grievances against the West or that the possibility of Communist revolutions has disappeared. By no means. A few peoples have still to pass through their anti-imperialist revolutions; unpopular vestiges of the old imperialism remain in some of the new nations; the new nations resent pressures on them to become the West's avowed military allies (this is true in India even after the attack by Red China) and any "meddling" with their foreign policies; and they deplore the billions spent on competitive armaments that they think might better be spent on economic growth, both in the old countries and the new. Most important, not a single underdeveloped nation has made the historic breakthrough to industrialism by using nontotalitarian methods, and not a single one of them has achieved assured stability. It may be that some of the anti-imperialist revolutions will still pass through a second phase, that some will yet merge with Communist revolution.

But the truth is that in the main the first and potentially most explosive phase of the anti-imperialist revolutions has passed into history, that the new nations have gone through their anti-Western revolutions and today still have closer material and spiritual ties to the West than to the Communist countries, that the former colonial peoples have given the West a second chance.

By and large, there is not only less revolutionary opportunity in the world than there was even a decade ago, but there is also a decline of revolutionary zeal in the Soviet Union itself.

The interests of the Soviet Union are increasingly on the side of the status quo. Historically, all revolutions have a way of slowing down with age. The Russians have now successfully passed through the drastic stages of their industrial revolution. As Barbara Ward has pointed out, Russia, on the basis of per capita income, is no longer one of the "poor" nations but has become one of the "rich" nations. Russian leaders are enormously proud of their vast new industrial equipment and technological developments, and like vested interests everywhere they put a high priority on protecting "their" properties and their positions. Russia's proliferating managers, engineers, technicians, and professional people have become accustomed to making important decisions in their work and are becoming more career-minded and status-conscious. The mass of Russians have been or are being transformed from peasants into urbanites, with rational standard-of-living values, who take their creature comforts (still modest by American standards) more and more for granted. In short, Russians are less and less interested in colorful revolutionary crusades, more and more interested in colorful personal living.

At bottom, it is this growing wealth and "bourgeoisization" of Russian society that so disturbs the Red Chinese, who are still desperately poor and have yet to make their industrial breakthrough. But if and when the Chinese pass over the industrial hump, will not they go the way of the Russians, will not their own revolutionary zeal abate?

Even had the old revolutionary enthusiasm persisted, the Communists today would be in no position to exploit it effectively, for they are beset by internal schism. (And schism, in turn, has had something to do with the cooling of revolutionary ardor.) History is disposing of Marxism in a way different from that predicted by Marxist theorists. The international proletarian brotherhood, the international proletarian society, is a myth. It does not exist. Communism has invariably taken a national form, and the various Communist nations do not present a solid phalanx vis-à-vis the non-Communist world. Communism was established in various countries at various times and in various ways. Communist countries differ from one another in their histories, cultures, conditions, interests, stages of revolutionary development, and degrees of Marxist "orthodoxy" and "revisionism." Today there are anti-Stalinist Communists, Stalinist Communists, Mao Communists, Gomulka Communists, and Tito Communists. Sino-Soviet relations fluctuate, but the rift between Communist Russia and Communist China may prove so serious that in world politics Russia, for the long run, will wind up closer to the West than to Red China.

As the decades roll by and deviations become revisionisms and revisionisms are piled one on another, will not Marxism become largely ritualistic, largely irrelevant as a drive to practical action?

Communism, of course, may be something much more than a mere historical instrument for the rapid transforming of preindustrial societies into industrial ones. It may be that Communist societies, as they industrialize, will not develop into variant middle-class societies. It may be, too, that Communist elites will know how to preserve their power structures and that the mass of people in Communist societies, because of uniform educations among other things, will sincerely believe that the collectivist society represents the superior way of life. Nevertheless, it is difficult to see how the pristine revolutionary evangelism can survive the growing affluence of industrialism.

Most important, no amount of revolutionary zeal could have overcome the stark fact that force is being banished from history, that the nuclear revolution is making international violence obsolete as an effective tool of politics. This will prevent the worldwide Apocalypse between "imperialist capitalism" and Communism. It is even making increasingly difficult any military intervention in behalf of revolutionary movements in non-Communist countries. The *pax atomica*, or coexistence, or stalemate, or hardened status quo, or freezing of actual power situations in all parts of the world—call it what you will—is already an almost accomplished fact, even though some Americans and the Mao Communists prefer to think otherwise, and even though it is never fully formalized by agreements. Not only do the two great nuclear powers (and even Red China) shy away from the big war, they also shy away from the little wars that might become the big one—witness Quemoy, the Himalaya border dispute, Laos, Vietnam, the Congo, Cuba, Berlin. Even the seemingly most reckless brinkmanship does not result in war. All attempts to preserve war as an institution that changes human affairs, all concepts of limited war, controlled war, graded and guided war, founder on the fear that any kind of war may eventuate in the big one, in nuclear holocaust.

The same fear that prevents anti-Communists from attempting a rollback of the Communists is likewise inhibiting the Communists from active military intervention where "objective conditions" seem ripe for social revolution. Most of the status quo is frozen. With both sides possessing nuclear weapons, the Communist Revolution will have no Bonaparte and the anti-Communists will need no Metternich.

Many forces, then, are moving to make the last half of our century one

79

of stability, but the clincher is the nuclear revolution in war. It may now become a frustrating, maddening, agonizing stability, but it will be stability of a kind, nevertheless.

It is possible, of course, that the last of the century will witness additional sweeping revolutions and a general nuclear war. There is no denying the explosive forces in the world. But do we not exaggerate them?

Some countries in Western Europe have not even yet escaped the possibility of Communist revolution. Communist parties in France and Italy are still strong. But these are not making such gains as would carry them to successful revolution; in neither country is there a revolutionary situation; neither party has a true revolutionary bent; both, like the older Socialist parties, have become mainly parties of protest within the existing system; their voting strength rises and falls depending upon the particular national political situation of the time.

It is possible that one or more of the important underdeveloped countries (say India or Brazil), failing to achieve the breakthrough to industrialism or even to accomplish a perceptible rise in mass living standards, may yet experience Communist revolution. But the truth is that there is probably less urgency to achieve industrialization in the underdeveloped societies than is often assumed. The picture we have of underdeveloped peoples panting for rapid industrialization is a bit exaggerated. There are popular demands for economic betterment, but the rise in living standards need not be sensational; a gradual advance will probably suffice in most cases. We are learning more about how to aid the underdeveloped societies effectively; and while we likely have a longer time than we thought we had, it is well to remember, on the other hand, that we do not have unlimited time.

The difficulties of the underdeveloped societies are complicated enormously by the population explosion. But the conscious tackling of the population problem is new; Japan has made great strides in curbing her population growth; Nasser's birth-control program in Egypt shows promise; and even the Chinese Communists, who like all Marxists theoretically denigrate Malthusianism as un-Marxist, seem to be successfully persuading the Chinese to use contraceptives. Even if population pressures alone produced wars in postprimitive cultures, a proposition of doubtful validity, such wars would be inhibited today by the nuclear deterrent, just as wars produced by other causes are being so inhibited.

Khrushchev may fall from power and his leadership and doctrine of peaceful coexistence be discredited, but this is not likely. Adjusting Marxist doctrine to the nuclear realities, dropping the inevitability-of-war dogma, will probably be regarded more and more as mere common sense. The nuclear revolution in war is infinitely bigger than Stalinism or Maoism. The likelihood is that the Khrushchev view will be sustained and in the end will prevail in all the Marxist world, because Marxist thinking emphasizes the decisive impact of technological change on society and politics, including war.

It is possible that nationalist and revolutionary evangelism in Red China will increase, that China's ancient ethnocentrism will combine with Marxist fervor to the extent of completely blinding China to the nuclear realities. But are not our fears excessive? It should be recalled that "orthodox" Marxists have regarded war between the capitalist and Marxist powers as "inevitable" not because of Communist aggressions but because of "capitalist-imperialist" aggressions. In her isolation, China is still much more suspicious of the West than is Russia. But is China today as militarily aggressive as we usually assume? Did not China, along with the United States, take pains to limit the Korean War? Did not China recoil from the brink in the Formosa Straits, in Vietnam, in Laos, in the Himalayas? What reason is there to suppose that China (heir to an essentially pragmatic cultural tradition), once she gets nuclear weapons, will be less prudent in their actual use than Russia and the United States have been? Would not more contacts with the West and an increase in affluence moderate rather than intensify revolutionary enthusiasm in China?[1]

Again, the American right wing, misreading world trends as favorable to Communism when on balance they are unfavorable, and ignoring the implications of the nuclear revolution in war, may come to power in the

1. By 1970, one of the most explosive situations was along the Sino-Soviet border, where it was the Russians and not the Chinese who were the more aggressive. The Soviets were repenetrating Sinkiang, and in frontier clashes with the Chinese they revealed an apparent urge to expand in Central Asia at the expense of China, whose aims and actual strength were still ambiguous. However, the Russians would proceed cautiously; they had a large stake in world stability; they probably would not risk an embroglio with China for fear of setting in motion a train of world-wide repercussions they could not control. Sometimes the Soviets seemed to flirt with inflicting a nuclear blitz on China, but, among other things, the knowledge that this would bring the United States and the Soviet Union to a face-to-face nuclear confrontation of unparalleled intensity operated to restrain them. The nuclear deterrents of the two superpowers moderate their relations with each other; they also moderate their relations with third powers. Although the United States has not relied sufficiently on a

United States and build a garrison state, pursue policies of aggressive unilateralism, wreck the United Nations, alienate America's allies, split the
West, reunify the Communist powers, give validity to the Stalinist and
Maoist interpretation of Marxism, and precipitate the Grand Apocalypse
of earlier Marxist prophecy. While this is a possibility, it is certainly not
a probability.

Then there is the chance of nuclear war by human miscalculation or human or technical accident or mistake. But there is much romanticism
about this; here is the stuff of which science fiction is made. This is possible, but not probable, and the technical controls over the nuclear deterrents are improving and tightening all the time.

Will not the chances of miscalculation, mistake, and accident multiply
as the nuclear powers proliferate? But is the number of nuclear powers
actually increasing? Britain seems more and more inclined to give up her
independent nuclear deterrent. Chancellor Adenauer is reputed to believe
that De Gaulle will eventually abandon his independent nuclear force because of the prohibitive costs. West Germany is debarred by international
treaties from having its own nuclear force. After France elected to build
a nuclear force, it was believed that other countries would follow suit—
Italy, Japan, India, and perhaps even Sweden and Switzerland, to protect
their traditional neutrality. But such was not the case. What is impressive
is not the alacrity of the countries to become nuclear powers but their reluctance to do so. Since the negotiation of the limited nuclear test-ban
agreement, it will be more difficult for additional countries to become
nuclear powers. And even if the nuclear club should grow, would not the
belated nuclear powers show the same restraint that Russia and the
United States have shown with respect to their nuclear deterrents, and for
the same reasons?

Never in history have war and world politics been so tightly in the grip
of such stupendous and rapidly changing technological developments as
they are today. Any predictions about the future must necessarily take
into account the possibilities of future technological breakthroughs. It is
possible that there will be a notable lag in the military technology of one
side or the other, thus inviting an attack by the power that has upset the

---

Russian-Chinese rivalry to check an "aggressive" China and maintain an Asian power balance, in the event Russia herself actually threatened a nuclear blitz against
China and jeopardized that balance, the United States would be compelled to resort to nuclear brinkmanship to protect the currently weaker power, China, and thus
encourage an Asian balance.

military balance. Again, some technological breakthrough, achieved by both sides, might for a time neutralize the possible use of cataclysmic weapons and restore war to a less awesome state. But it is not probable that defense devices will appreciably or for any length of time neutralize the nuclear armories. Hence the inhibiting stalemate produced by the balance of terror is likely to continue.

The catalogue of explosive possibilities in the world is indeed a long one. But the essential point is that they are merely possibilities and not probabilities. What is curious about the current situation is that we insist on overemphasizing the explosive *possibilities* and on enormously playing down and even ignoring the *probabilities*, which point strongly to increasing international stabilization.

Where, then, are we headed? It now seems clear that we are not headed for any monolithic world, Communist or American; or for a world divided into two monolithisms, the one Communist and the other American. Depolarization is taking place. Moscow is a declining influence in the Communist orbit. Washington is a declining influence in the non-Communist orbit. (The obstreperousness of De Gaulle and the growing inclination of all the NATO partners to renege on their NATO commitments, and even when they claim to want a multilateral nuclear deterrent to haggle over its cost and the granting of bases for its Polaris submarines, are strong indications that the Europeans essentially agree that the military threat of the Soviet Union and of Communism is passing, that American and Soviet nuclear might will operate as a mutual deterrent.) There has been a restoration of multi-diplomacy, and China and Western Europe may develop into such centers of power as to produce a revival of something akin to a multiple balance of power. At the same time, regionalism and even internationalism are growing.

Not monolithism but pluralism is the dominant political trend in the world politics of our time: depolarization in the East; depolarization in the West; the multiplication of neutralist states (which is not being arrested by China's foray into India); the emergence of scores and scores of new societies with fluid and mixed economies, pragmatic and varying combinations of indigenous collectivism, privatism, state enterprise, and welfarism; and marked differences even among Communist societies.

The first part of the century was one of revolutions, civil wars, total world wars. The second promises to be one of reconstruction, settling down, stabilization, adjusting the new to the old, constructing new societies with pragmatically mixed economies—and an epoch-making utilization of new sources of energy and conquest of outer space.

The constructive trends of the latter half of the century may continue under benign international arrangements which make beginnings in the control of armaments and the institutionalizing of peace and are progressively widened and implemented to provide for peaceful, desirable change. Much more likely they will continue under the awful threat of the nuclear terror, with all the economic and psychological cost that entails. But in either case, these trends will probably continue.

The major concern of the last decades of our century is likely to be not revolutionary and counterrevolutionary mobility, but rather the inhibiting immobility of a frozen status quo.

During the first half of the century men said: "Is there no escape from revolutions, civil wars, international wars?" By the end of the century men may well say: "Is there no escape from the imprisoning status quo?

How in the future will we settle the ultimate disputes of nations? What of justice and injustice? How ever break out of the status quo? How get desirable change? Will we find substitutes for the great change-makers of the past—international wars and the revolutions and counterrevolutions and civil wars which often spilled over into world affairs to produce international wars?[2]

In any event, it now appears that for several decades to come we shall face an essentially frozen international status quo, imposed in the final analysis by the restraining balance of nuclear terror.

---

2. One of the stock arguments against an effective international peace-keeping organization has been that in its efforts to preserve peace it might prevent change and freeze the status quo. Actually, the League of Nations used the plebiscite as one of its means to effectuate peaceful political change, and the United Nations has used it. After a few decades of immobility produced by the nuclear stalemate, an international collective-security organization might at last be welcomed for its very ability to make desirable changes in the international status quo, and to make them peacefully. This may ultimately be a factor in elevating the United Nations and other international agencies from a secondary to a primary place in world politics; but that time is not now.

II

American Politics

❧ Centralization and
the Open Society

The political folk mind in America is essentially Jeffersonian, and for a long time even those who benefited from the expansion of federal activities had an uneasy feeling that both democracy and liberty were threatened by federal centralization. The popular fears were exploited by local power elites, who got a better "deal" from their state governments than from Washington. Those opposed to enlarging the functions of the federal government reduced Jeffersonian doctrine to a shibboleth, and emphasized its *form* over its *substance*. The following article argues that the expansion of the federal government, far from impairing democracy and liberty, has actually widened both.

By the late 1960s a new movement was afoot to arrest federal centralization, encourage the states to assume more governmental functions, and funnel federal money to the states in block grants to be spent as the states themselves should determine. It was argued that since the states had by then modernized their state constitutions, reapportioned their state legislatures on a more democratic basis, and enfranchised their cities, they were in a better position to deal with modern problems than they had been formerly. Those who read the following article with understanding will be wary of these arguments. The traditional local power structures still have more influence in state governments than in the federal government. State politicians, officials, administrators, and judges still respond to fewer and narrower interests than do their federal counterparts. State administrators are still less well trained, less well supervised, less well audited, and less systematic in their methods than are federal administrators; their activities are still less exposed to the white heat of publicity; and a large number, even at this late date, are still products of the political spoils system. The mayors of the cities report that they still get a more sympathetic understanding of their problems from the federal government than from their state governments. And ironically, the booming cities in those Southern and Western states with fast-growing new industries have spawned a new type of successful urban politician, the right-winger, far more rigidly doctrinaire in defense of the new rich than the old Bourbon agrarian ever was in defense of rooted landed property. This new type of urban

Political Science Quarterly, LXXVI (June 1960).

87

politician is often sent by the cities to the legislature, and in any event he and his kind have more influence in state legislatures than they have in Congress.

Does the following article exaggerate the present-day pluralism in American life? Is it correct to say that "all parts of America are becoming more alike, but more alike in their diversity"? While it is true that industrialism in its earlier stages breaks down the old custom-bound parochialism of agragian life and creates multiplying diversities in all parts of a country, does not an advanced industrialism, a technocracy, move away from pluralism and in the direction of a stereotyped, other-directed, nonautonomous society? A later article in this book, "American Intellectuals and American Democracy," suggests the ways in which an advanced industrialism still creates new opportunities and diversities; and the article after that, "The Century of Technocracy," examines the counter-trends and Orwellian threats produced by technocracy. Both should be read together and then related to the following essay on governmental centralization.

Of one thing we may be sure: If the full potential of technocracy's benefits are to be realized and spread widely over all of America's classes and groups, we shall have to look, in so far as the political process is relevant, more to the federal government than to the states. Again: In so far as the abuses and threats of technocracy are amenable to political and judicial processes, we shall likewise have to look more to the federal government than to the states. And in implementing measures today it is urgent that all governments—local, state, and federal—constantly keep in mind the need to foster social units with which individuals may personally identify themselves in a sense of community.

■ IN 1952, the theme of Dwight Eisenhower's presidential campaign, at least in domestic politics, was the reawakening of the states, the restoration of their traditional place in the federal system, a return to the values and practices of decentralized federalism. Even scholars took another look at the federal system, and some became moderately optimistic about the prospects of revitalizing the states.

Among those who shared this optimism was the late Leonard D. White, one of our most distinguished scholars in the field of public administration. In a series of lectures at Louisiana State University, published in book form in early 1953 under the title *The States and the Nation*, Professor White set forth the grounds for believing that the federal government would relinquish some of its tax sources to the states, among others the gasoline tax, and that the states themselves would also explore new sources of tax revenue.[1]

Professor White believed that the states would probably revitalize themselves in other ways. First, they would relinquish some federal grants-in-aid and assume independently some of the services they now render jointly with the federal government. Secondly, the states would experiment more widely with the interstate compact. Thirdly, they would make a much fuller use of their own powers in two general ways—they would enlarge their traditional services and keep these abreast of the great advances being made in criminology, penology, mental health, and education; and they would resume their old role as laboratories of experimentation by taking on new functions such as adult education and training, government aid to superior students, health insurance, possibly even disability insurance.

This emphasis in 1952–1953 on the bright future of the states was, of course, a reaction from the enormous growth of federal power during the 1930s and 1940s. As late as the turn of the century, the federal government had affected the national economy only through its simple excise taxes, its protective tariff, its uniform currency, its marketing of federal bonds, the limited functions of the national banking system, and an inconclusive regulation of interstate carriers. By 1950, however, the activities of the federal government had come to touch vitally every aspect of American life. The old and original powers of the federal government had been prodigiously expanded by new demands in foreign affairs and national defense. And the New Deal and the Fair Deal, building on the earlier New Nationalism and New Freedom, had charged the federal government with vast new functions so as to assure a smooth functioning of the American economy. By 1950, American constitutional theory and practice had come to accept this and American opinion overwhelmingly to expect it.

In 1960, it looks as if the anticipations of a revived federalism in its old forms were unfounded. Aside from tidelands oil, there has been no significant return of functions or tax sources to the states. Since 1953, it is true, there have been demands for an enlargement of old state functions and for the initiation of new state functions, but most of the states fail to respond to the demands. At the same time, the federal government has been taking on and expanding functions hitherto regarded as largely

1. Leonard D. White, *The States and the Nation* (Baton Rouge: Louisiana State University Press, 1953).

89

state or private matters—road-building, welfare, housing, slum-clearance, urban renewal, individual and public health, hospitals, education, and so forth. The federal government has been concerning itself more and more with these activities, either directly, or indirectly through grants-in-aid. Older grants-in-aid are expanding, new grants-in-aid are being adopted, and federal money appropriations for grants-in-aid are increasing enormously. In 1901, they amounted to less than $3,000,000. In 1931, they amounted to about $220,000,000. In 1941, they rose to $615,000,000. By 1951, they had expanded to $2,280,000,000. Today, they are running close to $7,000,000,000, counting the so-called highway trust fund.[2]

In effect, the centralization-decentralization issue has now shifted from the American *economy* to the American *society*. A new rationale to justify a new expansion of federal power is developing. More and more it is being said that our society is national, that as a society we are becoming more interrelated and interdependent, that we are a more mobile people than ever before, that health and education in one state affect health and education in all the states and in the nation.

At the turn of the century it would have taken a bold man to declare that since our *economy* was national, the government most concerned with that economy must be the national government and not the state. Today this is a commonplace. Now it is being said that because our *society* is national, the government most concerned with that society must be the national government and not the state. Today this is an advanced attitude, but tomorrow it may be a commonplace.

From many sides are coming demands for enlarged and new government services. Here are a few straws in the wind. In a recent series of articles on urban renewal, *The Christian Science Monitor* estimated that one out of every four city dwellers in the United States lives in a slum.[3] A short time ago the New York State Department of Labor, commenting on the rising costs of health services, reported that today two out of every three dollars spent for medical and hospital services in the United States are still borne by the individual, that only one dollar in every three is covered by some kind of insurance.[4] The National Education As-

2. White, *op. cit.*, p. 18. An address by George C. S. Benson, in *Governmental Research Association Reporter*, 11, No. 1, 1–7.

3. See editorial in *The Christian Science Monitor*, July 13, 1959, summarizing that newspaper's series of articles during June and July on slum conditions and urban renewal.

4. The New York *Times*, Sunday, May 10, 1959.

sociation is insisting that our public schools need an additional eight billion dollars each year if American education is to be what it should be quantitatively and qualitatively, and it strongly implies that all estimates of federal aid up to this time have been trifling in comparison with existing need.[5] From all sides come reports of skyrocketing costs for college and professional training. There is the much-publicized article of Benjamin Fine, who estimates that about two hundred thousand of our most gifted high school graduates each year cannot go on for college or technical or professional training for lack of financial help.[6] And the nation is reacting to all this differently from the way it would have formerly. Yesterday people would have said: "But advanced education is for the individual to provide for himself." Today people increasingly say: "What a waste of national brain power!"

These new demands are coming out of new conditions. What are these conditions? Rapidly increasing population. Continuing and accelerating shifts of population to the urban areas. Disproportionately rising costs in medical services, hospital services, and education. The expansion of the private sector of our economy, and the realization that as the private sector expands so also must the public sector—for otherwise we face grave imbalances. (For instance, when the private sector builds more automobiles, the public sector must build more roads.) The growing recognition that goods and services produced by the public sector are not just luxuries or consumer goods; that they are productive wealth, used to make more and better goods and services. (Better schools and better health facilities produce more effective workers, managers, scientists, technicians, technologists.) The deepening realization that we must keep ahead of the Soviet Union in industrial and military technology. The growing consciousness that desegregation is a vital ideological weapon in our world struggle to win the uncommitted peoples of Asia, the Middle East, Africa, and Latin America, and that we can no longer regard this question as a domestic and local one. In short, crisis is the health of centralization, and we are in a continuing crisis.

5. Associated Press account appearing in scores of American daily newspapers on June 30, 1959, summarizing the report of the Educational Policies Commission of the National Education Association made to the Association's convention on June 29.

6. Benjamin Fine, "College Tuition Soaring Higher," an article syndicated by the North American Newspaper Alliance, appearing in scores of American newspapers in June 1959, and interpreting the report of Dr. Ernest V. Hollis, Division of Higher Education, United States Office of Education, on mounting costs of college education.

II

Will the states do their part in meeting the nation's growing expectations? Have the optimistic predictions about the future of the states been borne out by subsequent developments? The evidence seems to be that they have not, that the states will not sufficiently revitalize themselves, that the trend to federal centralization will continue.

True, all states are making some progress; and a few states have taken constructive, even heroic, measures to tap water supplies, find new tax sources, provide regional planning, and reapportion their legislative seats. But what was the record of most of the legislatures meeting during 1959? It was largely a record of economy, retrenchment, and the slashing of budgets submitted by the executives. The truth is that most of the states are barely able to keep the old services abreast of increasing costs, increasing population, and the shifts of population to the cities.[7]

Few of the things envisaged by Professor White have taken place. Have the states relinquished any federal grants-in-aid? Not at all. Not a single grant-in-aid has been relinquished. Here is the typical story. First, a grant-in-aid is made so as to get a necessary or desirable government service started and to soften opposition to it in the states. Then it becomes a going concern, vested interests are created, the controversial becomes customary, and the opposition vanishes. Therefore, grants-in-aid do not diminish; instead, they multiply.

What of the interstate compact? It still remains largely a potentiality rather than an actuality, although there is an encouraging tendency by the states to experiment with it more widely. The truth seems to be that the interstate compact works most successfully in the noncontroversial, that is, the relatively unimportant, areas of activitiy such as the regional educational councils, the return of parolees, and so forth. In the controversial areas—such as electric power, tapping water supplies, preventing water pollution, conserving soil and other natural resources—it has been less successful, although longer and wider experience may bring greater achievement. At the present time what is most impressive, considering the many possibilities for the use of interstate compacts, is this: how few are attempted; of the few attempted how many founder in the proc-

7. For what did the state governors and budget directors ask their legislatures in 1959? A summary, state by state, is in *State Government*, the Council of State Governments, 32, No. 2 (Spring 1959), 78–103. And what did the legislatures do? A summary of actions by the state legislatures, state by state, is in *ibid.*, No. 3 (Summer 1959), pp. 199–212.

ess of negotiation and ratification; of the very few that materialize how prolonged and difficult is the process of negotiation and ratification. Proposed interstate compacts must be watched closely, for they sometimes contain built-in devices for local vetoes, disguises for obstruction.[8] It is instructive that one of the reasons for calling the Constitutional Convention of 1787, to form a stronger general government, was the failure of the Potomac River states to conclude an interstate compact. In the light of the total situation and the many opportunities for employing this device, the use of the interstate compact is still negligible.

Are the states making an adequately fuller use of their own powers? They are not, emphatically not. This is so, chiefly because of the realities of group politics in the states. It is true that the groups that press for the expansion of government services to meet new conditions can be found in all states, but it is only in the industrial states that they can exert much influence. Generally speaking, they are more numerous and better organized on the national than on the state level. On the state level they frequently run into structural barriers erected by nineteenth-century horse-and-buggy constitutions. Some of these are so long and involved as to be in effect codes of law rather than constitutions. They place restrictions on the state's taxing power, borrowing power, spending power, and on the activities of the cities. Above all they provide for flagrant malapportionment of the legislatures.

Malapportionment of state legislatures is by now an old and a familiar story: how the rural areas are overrepresented and the urban areas underrepresented; how as little as 30 or 25 or 20 per cent or even less of a state's population frequently makes a majority in the legislature; how one vote in a rural area often equals one-hundred or two-hundred or even three-hundred votes in an urban area.

What is not so well known is the extent of this malapportionment, how general it is, how flagrant, how today some of our states are as undemocratic as Great Britain was before the Reform Bill of 1832.[9]

The consequences of all this are even less well known. Overrepresentation of the rural areas means not so much the rule of the rural folk as it does the rule of the rural politicians, who largely reflect the interests and

8. William E. Leuchtenburg, *Flood Control Politics: The Connecticut River Valley 1927–1950* (Cambridge, Mass.: Harvard University Press, 1953), chapter 2; Arthur W. Macmahon, ed., *Federalism: Mature and Emergent* (Garden City, N.Y.: Doubleday, 1955), pp. 344–347.

9. Manning J. Dauer and Robert G. Kelsay, "Unrepresentative States," *National Municipal Review*, XLIV, No. 11 (December 1955), 571–575.

values of the county rings, the large landowners, the small-scale and local-minded business men of the county-seat towns, and certain corporate businesses (themselves located in the cities) which would rather deal with a legislative oligarchy than a broadly representative and democratic legislature.

It is the legislatures dominated by rural politicians that favor the unrealistic proposals to amend the United States Constitution so as to limit the federal income tax, particularly in the higher income brackets. It is these legislatures which spawn the so-called "right-to-work" laws and other legislation designed to hamper the legitimate aspirations of organized labor. It is these legislatures which have a peculiarly tender regard for the small loan companies. (Recently, a bill sponsored by no one except the small loan companies was passed by the legislature of Florida; the "aye" votes in the house represented constituencies totaling 600,000 persons, while the "no" votes, which failed to stop the bill, represented constituencies totaling over two million people.[10])

Domination of the legislatures by rotten-borough rural politicians is especially dangerous in the one-party states, where these politicians become still more inbred and form a ruling clique unchecked by even a rival party clique. Nor does this situation prevail only in the South.[11] About one half of all the states can be called one-party states, at least so far as state matters, especially the legislatures, are concerned.

These conditions are all roadblocks on the way to making the changes that would have to be made before the states could embark on that fuller use of their powers that was predicted by Professor White. Let us list some of these necessary changes: constitutional revision; genuine reapportionment; the exploration of new sources of taxation; the establishment of adequate civil service and merit systems in the states so that they could compete with private enterprise and the federal government for administrative personnel; emancipation of the cities from neglect and exploitation.

Thus the formidable task in most of the states is a twofold one. First, the structural barriers must be removed. Then the constructive battle must be fought to expand old services and introduce new ones.

10. An unpublished manuscript on the effects of malapportionment of legislatures on specific legislation, by Manning J. Dauer, Department of Political Science, University of Florida.

11. V. O. Key, *American State Politics: An introduction* (New York, Alfred A. Knopf, 1956), chap. 3.

The chances are that it will be another case of "too little and too late" since in the meantime the federal government will be moving in with new or expanded services which are given either directly, or indirectly through additional grants-in-aid. In fact, we may be nearer than many of us suspect to another great spurt in federal expansion. If the present Congress were not checked by presidential veto, we would be in the midst of it right now. After January 1961, a Mr. Veto may no longer be in the White House.

III

Yet we must beware of exaggeration. The federal system is moving in the direction of greater federal power, but it is not becoming extinct. Institutions are tenacious things, and they are transformed slowly. Local governments for experimentation, adaptation, apprenticeship in political leadership, and training for democratic citizenship will continue to function.

The states will continue to create and maintain counties and cities for local government. The states will continue to use the counties as agencies for the administration of state matters, though not as much as formerly, for the state today tends to administer more of its own functions directly and not indirectly. The states will continue to give their own grants-in-aid to the counties—so that the state will increasingly spend federal money it does not collect and collect state money it does not spend. The states will continue to exercise many powers independently of the federal government. Even where the federal government starts exercising a power formerly exercised exclusively by the state, the federal government does not always pre-empt the field, and thus co-operative federalism comes to operate in a wider way. (This whole subject of co-operative federalism is a large and intricate one in itself.) Finally, the state will continue to administer the joint federal-state projects under the federal government's expanding grants-in-aid programs.

Actually, in this century the state has been increasing its powers. State government does more than it did at the turn of the century. Some of these added powers have been taken from the counties, and some of them are new powers never before exercised by any government in the United States. However, the federal government, as we have seen, has taken on more powers, too. Some of these added federal powers have been taken from the states, but some of them are new powers never before exercised by any of our American units of government. The federal government

95

has been increasing its power at a more rapid rate than has the state, and its expanding powers are more significant and pervasive, more ramifying in their effects. Thus while the state has increased in power absolutely, it has declined relatively.

By now, we ought to face bluntly the implications of federal centralization for our open society. About this, there is profound and widespread confusion, even among our "liberals" and "progressives." The "conventional wisdom" places a high premium on the values of decentralized federalism, regarding it as the source of experimentation, adaptation to local variations, apprenticeship for political leadership, training in democratic citizenship, the very existence of democracy, the very preservation of liberty. Does this square with the facts?

J. K. Galbraith in *The Affluent Society* has shown how the conventional wisdom in economics—the discrepancy between our old stereotyped concepts of the traditional free market and the realities of today—clouds our thinking and actions in economic matters. Is it not high time to analyze the conventional wisdom in politics and to inquire whether there is not a similar discrepancy between the stereotyped concepts of the old decentralized federalism and the political realities of today? Is not the conventional wisdom in economics linked with that in politics? Do not our anachronistic, ritualistic, and mythological notions about political decentralization help perpetuate our anachronistic notions about the "reality" of the free market today? Are not our horse-and-buggy state governments among the most formidable barriers to our meeting present-day economic and social needs? Do not people who recognize today's limitations on the free market but at the same time cling to the primacy of the local governments in effect nullify their insights into economics with a superstitious political faith?

It seems to me that an examination of the realities of contemporary politics would go far to show that federal centralization means wider economic and social fulfillment for the individual and the nation, and at the same time means not less democracy but more democracy, not less liberty but more liberty.

IV

Even if the federal system were becoming extinct—which it is not—there would still be local governments. It is a rather naive and parochial notion that we must have a federal system in order to have local governments with administrative and even discretionary powers. Even governments that are not federal at all—that are national in character—decentralize and allow for considerable creativity in local government. Most of the governments of Western Europe are national, not federal, but they allow for local governments. These local governments provide apprenticeships for political leadership. Adenauer was mayor of Cologne. The Chamberlains saw service in the municipality of Birmingham. Attlee learned his politics in the London County Council. Many of the members of the French Chamber of Deputies served their towns and cities as mayors. If our American cities succeed in freeing themselves from some of the unreasonable restraints of the states, this might mean a further weakening of the traditional federal system, but it would surely result in a great energizing of local government!

In many ways experimentation in state and local governments is on the increase, particularly in technical and administrative matters. Home rule for the cities, where it exists, allows a wider scope for experimentation. The federal government, in gathering, organizing, and disseminating data about state and local governments and in insisting that the states, under the grants-in-aid programs, keep better records and organize data, is actually stimulating experimentation and adaptation to local needs.

We must guard against an unwieldy federal bureaucracy, but we must also remember that federal administration and federal administrators are not the evil things they are often said to be. Federal administrators are in general better trained, better protected by civil service, and better audited financially than are state administrators. We should keep in mind that because a power is federalized it does not follow that its administration will be centralized in Washington. Federal administration may be regionalized, and indeed some of it is already regionalized. Even so, federal administrators do administer over a wider geographical area, and therefore minister to a larger number of groups, interests, and values, and must be responsive to them. The people can influence federal administrators through their local congressmen and senators. Practical politicians frequently observe that people today have a greater feeling of intimacy with the federal government than they have with their own state and local governments, that people know their local congressman better than they do their local state legislator, that they bring him not only federal requests but many state and local requests.[12]

12. Emanuel Celler, *You Never Leave Brooklyn* (New York; J. Day Co., 1953), p. 266. Congressman Celler's experiences have been largely confined to the concentrated

Now, it is just as important to resist improper pressures as it is to respond to legitimate and democratic ones. Federal officials and administrators, since they minister to a wider area and to more groups, interests, and values, are in a better position to resist improper pressures. If they offend some groups, they may compensate by increasing their popularity with other groups. Federal administrators are farther removed from particular local prejudices; they get a wider perspective; it is harder to convince them that a handful of people represent the whole universe. Usually, what we mean by an official's "growth" is the result of his transferring from a position of narrower horizons to one of wider horizons.

True, there recently have been disclosures of improprieties in some of the federal commissions in Washington—but that is just the point: in the federal government, because of the clash of more diverse and articulate interests, the federal commissions get the white heat of publicity, whereas the work of the state commissions all too often passes unnoticed. As a result of the recent disclosures, a great hue and cry has arisen to put the federal commissions beyond all pressures, to put them above and beyond the battle. In a democracy, this is dangerous nonsense. In a democracy, what we must avoid is secret and undue influence by some groups and interests; what we must guarantee is open access by all groups and interests.

There is a tendency to look back to the nineteenth century, when the state played a larger part in our federal system than it does today, and to say: "That was the day of real democracy, but now with all this growth of federal power, democracy is dying out." Now, it is romantic to think that we can return to an agrarian past. It is also romantic to think of that agrarian past as democratic. We look back and idealize the past. We visualize people in their rural neighborhoods and villages, in their town halls and county courthouses, all interested in politics, all participating in politics. Actually, politics in those days was largely a matter for professional politicians, for tight little caucuses and conventions, for county rings, for machines and bosses, for local bigwigs and notabilities. Politics was largely a matter for native, white, Protestant males, especially planters, big farmers, and the prosperous business and professional classes. Actually, nineteenth-century democracy was a limited and restricted democracy, a mere preface to our wider democracy of today.

---

metropolitan areas, but they tally with my own long-time observations of politics in Indiana and Florida and with the experiences of many practical politicians in those states with whom I have talked.

Today, our democracy is mass democracy, metropolitan democracy, melting-pot democracy, group democracy. Politics centers less on caucus, convention, courthouse, town hall, and state capitol. People learn, discuss, and operate in politics more through their functional groups—their economic, social, occupational, professional, business, labor, farm, ethnic, and religious organizations. Today, there are more groups; they are better organized; they are organized nationally. The newer, popular, and numerically larger groups have learned how to use their national organizations on the federal level to bypass the local traditional power structures, the local planter-lawyer-doctor-banker-realtor-insurance-business elites, to achieve a wider influence in public affairs and a larger respect for popular interests. This has not resulted in neglect of the interests of the traditional local elites—they still have an enormous influence—but it has resulted in a better and more democratic balancing of interests.

For people are still interested in local politics, and through their organizations they discuss and participate in local politics. Today, it is at the PTA meeting, the teachers' meeting, the League of Women Voters meeting, the woman's club meeting, the Junior Chamber of Commerce meeting, the civic club meeting, the union meeting, and the farm organization meeting that people learn about and discuss local politics. Thus today, more people, more different kinds of people, participate in local politics than ever attended the old courthouse rallies, which were largely composed of the politicians, the officeholders, their families, their friends, and those who wanted political jobs. Today, even in semirural states, widely representative community organizations, like the local and state PTA and the local and state League of Women Voters, are playing a relatively larger part in politics, and the old county rings are playing a relatively smaller part.

In general, the most undemocratic areas in America are the small rural counties, the very ones vastly overrepresented in the legislatures, for these counties usually have the smallest number of diversified and organized groups, and the professional politicians, with the voters atomized and less aware, can be less responsive to needs and more free to operate on their own, to serve their own interests at variance with those of their rural constituents. Today it is the heavily rural counties, where the traditional local elites still predominate, which continue to operate most like the way the "democracy" of the nineteenth century operated.

People who look back to the past emphasize "the freedom of the states." People who would come to grips with present-day realities em-

124988

EMORY AND HENRY LIBRARY

phasize freedom of association, the importance of group organization, and fair representation on all levels of government.

Last, and most important, what of liberty? The "conventional wisdom" in America has it that liberty is nurtured by local government and dries up when government becomes far away and remote. This does not square with the realities. Most tyrannies in history, and some of the most onerous, have been local. Indeed, a central trend of Western civilization has been the breakdown of the rule of local feudal lords and local magnates and the development of the national democratic state, which has done much to free individuals from local shackles.

The most frequent violations of constitutional process and civil liberties in the United States have been in the states and localities. Let us mention a few examples: the shoddy treatment of Negroes, aliens, and migratory laborers in many states; the Lusk gag laws of 1919 in New York state and the expulsion of the Socialist assemblymen from the New York state legislature in the same year; the control of Indiana by the Ku Klux Klan in the mid-1920s; the Huey Long dictatorship in Louisiana; the existence of prohibition in many states long before and long after prohibition on the national scene. Traditional decentralizers are in the habit of saying that prohibition in Kansas is all right because it is an adjustment to a local variation, but that prohibition on the national scene is a mistake. Are not a person's liberties as much impaired by prohibition from Topeka as by prohibition from Washington?

Think back, too, over our judicial history. Have not most of the judicial miscarriages been in the state courts, and not in the federal courts? There would be more of these were it not for the due-process-of-law clause and the equal-protection clause of the Fourteenth Amendment, and for judicial review in the federal courts by federal judges removed from local prejudices, sustained by a broad national opinion, and taking the large and long view.

In historical perspective, periods of extremism and hysteria on the national scene have been infrequent, short-lived, and relatively mild. Three of these stand out. One was the "black cockade" reaction of the John Adams administration, which gave us the Alien and Sedition Laws. Another was the Palmer Red Hunts of 1919. The third was McCarthyism and the spate of congressional investigations into subversion, without due regard for civil liberties, in the early and middle 1950s during and following the Korean War. All of these proved to be abortive. Witch-hunting on the national scene created a furor and collapsed several years ago, but it continues on the state level, where little McCarthys in some of our state legislatures have set up committees which are currently investigating subversion, un-Americanism, socialism, communism, the NAACP, immorality, and sundry other things. It goes without saying that many of these investigations are being conducted with no nice respect for personal civil liberties. On the national scene these would produce furious indignation and vigorous countermeasures, but in the states there is apathy.[13]

Even during the national trauma of the Civil War, the conduct of President Lincoln was moderate compared to the high-handed methods of a Governor Oliver P. Morton in Indiana.

There are several reasons why liberty is better safeguarded in the nation than in the state. The chances are greater of having independent citizens and newspapers to speak out against passion and injustice. It is harder for a single group or faction to get control of the government. It is more difficult to form a majority, and even more difficult to form an opinionated majority. Within the wide scope of the nation there are many more classes, groups, interests, and values which check, restrain, and counterbalance one another. All of this is stated in *The Federalist*, and it is as true now as when it was first written.

Is it not a curious paradox that today, when we are much more national than we were in 1789, these truths in *The Federalist* have almost ceased to be expressed, that they are all but drowned out in the steady drumfire of expressions of fear of federal centralization? Why do those who favor federal centralization today put their case almost exclusively in terms of concrete measures and refuse to put it in a wide and philosophic setting as the centralizers of 1789 did so admirably and successfully?

There are more groups today than there were in the early days of the Republic; they are better organized; they are more nationally organized. These groups help check big business; they help check big government; they help check one another. The group process is coming to be recognized as another of our guarantees of liberty along with the written Constitution, judicial review, separation of powers, and federalism. The

13. For the activities of one of these legislative investigating committees, the Florida committee probing into the NAACP and subversion in that state, see the files of the Tampa *Tribune*, Tampa, Florida, from mid-October 1958 through March 1959. The committee got sidetracked into an intensive inquest of subversion and immorality at the University of Florida. Not a single leader or newspaper in the state protested or even questioned the investigation or its methods.

states have declined somewhat and the federal system has become somewhat weaker, but the group checks have become stronger. Liberty is better safeguarded in those areas where there are many and well-organized groups than in the areas where they are fewer in number and less organized.

In the earlier days of the Republic, the states thundered at the government in Washington, and legislatures often instructed United States Senators out of their seats. Those days are gone. Today, it is the groups that restrain government and our political parties, and thus play a vital role in the processes of both democracy and liberty. Fortunately our two major parties in America are center parties; neither runs to extremes. But when the Republicans are in power in Washington and incline too far to the right, they are checked not so much by the states as by the activities of the nationally organized liberal groups. And when the Democrats are in power in Washington and incline too far to the left, they are checked not so much by the states as by the activities of the nationally organized conservative groups.

What happens when a national organization gets too ambitious and overreaching? Then other groups tend to ally themselves against it. And it is important to note that if government is necessary to regulate such a group, it is the federal government which must do it, for our leading national organizations are now too large for effective state regulation. Also, any hidden collusion of groups would ultimately bring resistance of other groups and intervention by the federal government.

Are the rank-and-file members of America's mass-group organizations exploited by their own leaders, their own elites? This and other aspects of our increasingly corporate and group society (particularly the spiritual and personality aspects) are beyond the scope of this discussion. However, it should be said in passing that it is a safe guide to assume that farmers, wage workers, and Negroes are better judges of their own interests and leaderships than are outside elites, and it seems indisputable that the individual members of mass groups now have better economic conditions, enjoy wider civil liberties, and exercise more political influence than they did in the days when they were unorganized and atomized.

All parts of America are becoming more alike, but more alike in their diversity. Formerly, the diversity of America was geographical: diversity expressed itself in differences from locality to locality and from section to section. Today, as industrialization and urbanization spread, every community is becoming more diversified, coming to contain within itself a larger number of different groups and values. This new pluralism, found on all levels but particularly on the national level, will do much to offset the relative decline of the states, prevent the dangers of federal centralization, and preserve and extend both democracy and liberty.

&2 The Growth of the Presidency

Since this article was published, seats in the national House of Representatives have been reapportioned on a more democratic basis, but minority groups in the big cities still look to the President for their popular tribune, and they are at sea when he does not assume that role. Also since this article appeared, the Vietnam war has sparked agitations to put more restraints on the President's vast powers in foreign affairs, but long tradition is against additional checks on the President in this field, and today the arguments against them are stronger than ever—with America the leading power in world affairs, with international events moving much more rapidly than in the past, and with the general recognition that the President's hands should not be tied in the event of a sudden and dangerous crisis.

In no aspect of his office does an American President think of himself as having too much power. The things that impress him are the obstacles he faces in meeting his responsibilities and the expectations of the public. Every President bemoans the lack of liaison machinery between Executive and Congress. Again, in this day of mass democracy, a President must communicate more often and more intimately with the public. Most Presidents, however, find that this is not easy to do in an effective way, despite the mass media at their disposal. President Franklin Roosevelt developed this art to a high degree, and in his "fireside chats" he cut through the complexities of his problems and in simplified—but not oversimplified—fashion explained them in language the people could understand. No President since FDR has had this capacity to a comparable degree, not even Kennedy, although it now appears that Nixon has markedly developed it.

Contemporary Presidents are beset by pitfalls that did not exist in the nineteenth century. One is the ubiquity and treachery of the polls which test current public opinion. Every President is tempted to follow the polls, but one who does so may find that he has been deceived. President Lyndon Johnson put too much confidence in these polls in the days his administration was escalating the Vietnam war, only to find that when the war became prolonged and the costs mounted, public support for it ebbed away. On the other hand, what if FDR had been overinfluenced by the overwhelming isolationist sentiment among Americans in the years the Axis powers were riding high, and in

Current History, XXXIX (October 1960).

consequence refrained from alerting the country to the Nazi danger and from making preparations to meet it?

Again, Presidents seem able to "size up" their fellow politicans better than they can estimate the personalities and counsel of their trained technicians, specialists, and braintrusters, necessary advisers of recent origin. With respect to the Bay of Pigs, President Kennedy always regretted not subjecting the advice of his "experts" to a more critical political examination. And did not the computer and the computer-minded help lure President Johnson into the Vietnam quagmire?

Today, too, a President must be particularly wary of making important decisions on information too much screened by his aides. He must also be on the alert lest he find himself bound by a *fait accompli* resulting from a series of decisions made by administrators in the lower echelons. In order to keep control of his administration, a President today must either be peculiarly perceptive and intuitive about what goes on inside the administrative hierarchy or spend an inordinate amount of time "riding herd" on the bureaucracy.

A sense of history is the most important single virtue a President may have; this rises above all institutional arrangements and all other personal characteristics.

■ AN AMERICAN PRESIDENT performs many roles; he performs all of them every day; and his effectiveness in one influences his effectiveness in the others. All of the presidential roles have been tremendously widened and deepened in the twentieth century.

The President is Chief of State, and today he is expected to take the leading part in far more ceremonials, ranging from the most elevated to the most folksy, than ever before in our history.

The President is Chief Executive and Chief Administrator, and today he presides over a vast and sprawling federal administrative establishment which makes the federal establishment of the turn of the century look picayune indeed.

Flowing from his roles as chief of state, chief executive, and chief administrator, the President is chief ameliorator of disaster and budding crisis. Let natural disaster strike, and he mobilizes the relief agencies; let labor-management or business dislocation threaten, and he is the first to take measures to avert a crisis. In the twentieth century a natural disaster affects more people, and crises of all kinds are much more common than they were in the nineteenth century.

During this century, the President has become chief legislator. In spite of the separation of governmental powers, he must formulate and fight for a legislative program in Congress. Indeed, today a "strong" President is one who can get his program through Congress; and when Congress adjourns without having passed much of a President's program, it is the President who is blamed, and not Congress.

The President is party leader, and he must actively take and keep control of his party. Since the President has now become chief legislator, his leadership activities within his party have greatly increased, for party leadership is one of the important weapons he uses in fighting for his program in Congress.

The President is Commander in Chief of the Armed Services in both war and peace, and during the United States' participation in the two total wars of this century, the President was given an awesome authority over the nation's economy and society. Today, with national defense continuously to the fore, the President makes decisions and recommendations in this field every day which affect in vital ways almost all aspects of American life.

The President is chief director of foreign affairs, and today American responsibilities in the world have expanded enormously. Presidential decisions and recommendations in foreign policy affect defense, and both foreign policy and defense have become central to American life; in the main all other questions and issues have become derivative.

The President is chief spokesman of the nation, the voice of the people, the great teacher. The White House has become the podium and sounding board of the country. In times of immediate excitement or crisis, the President must take to the radio and television to speak to and for the nation, and increasingly the President uses all of the mass media to mold long-time opinion, attitudes, and trends.

Until this century, Congress was considered to be the leading branch of the American government, as the long list of impotent, weak, dull, or stuffy Presidents who followed Lincoln and continued to Theodore Roosevelt well illustrates. Both Walter Bagehot, writing in the late 1860s, and Woodrow Wilson, writing his *Congressional Government* in 1884, attributed the confusion in American government to the ascendancy of Congress. True, prior to the twentieth century there had been foretastes of the President's later strength. President Jefferson was an active party leader and "embodied himself in the House of Representatives." Both Jackson and Polk were vigorous executives; both cracked the whip over Congress in fighting for legislative programs; both were strong party leaders. And Lincoln, during the Civil War, carried the presidency to breath-taking heights. But these were exceptions. Most Presidents until the twentieth century were "constitutional" Presidents and allowed Congress to take the lead, sometimes even in foreign affairs.

The great expansion in all of the President's roles has taken place in this century. Let us narrate some of the high points in this development and summarize the basic reasons for this trend.

Theodore Roosevelt, coming to office in 1901, was the first *modern* President. During his administration, every one of the roles of the presidency was widened and invigorated. He openly confessed to a belief in "the strenuous presidency"; he believed that it was not only the President's right but his duty "to do anything that the needs of the Nation demanded unless such action was forbidden by the Constitution or the laws." Roosevelt used the White House as a pulpit from which to preach the virtues of national strength and social justice. He dramatized the presidency; and his smallest activities, and those of his family, hit the headlines of the newspapers almost every day. It was in his time that the work of the presidency so increased that the President could no longer conduct it from the White House itself; during his working hours the President moved from the White House to the executive offices which were built on the White House grounds, and the staff of the President was enlarged.

Under Taft, much progressive legislation was enacted, but this was not because of strong presidential leadership in Congress but because of a remarkable bipartisan alliance of progressive Roosevelt Republicans and progressive Bryan Democrats in Congress. President Taft took a relaxed view of his office, belittled Roosevelt's notion of a strenuous presidency, and emphasized separation of powers and the "constitutional" role of the President. But as the century wore on, it became clear that it was Roosevelt's view, and not Taft's, which was to prevail.

Woodrow Wilson, like Roosevelt, contributed vastly to the growing strength of the presidency. Wilson was a confirmed believer in party government; he thought of himself as a kind of prime minister, with his party and himself as party leader responsible for a legislative program in Congress. The enactment of a remarkably coherent body of reform, known as the New Freedom, during the first Wilson administration, is probably the most notable example in American history of directed party unity and leadership.

Wilson was the first American President to become a real world leader, and much of his difficulty came from the novelty of America's first full-scale participation in world affairs. During the war, Congress gave Wilson immense powers over the American economy and social order, powers in many respects even greater than those exercised by Lincoln during the Civil War.

Under Harding, Congress gave the President control over the federal administrative budget. Until this time, every federal administrative agency made its own budget estimates for the following year and gave them to the House Committee on Appropriations and the Senate Committee on Appropriations. In 1921, Congress directed that henceforth the budget recommendations for every federal administrative agency, civil and military, must come from the President and from no other source. Detailed estimates were to be gathered and co-ordinated by a Director of the Budget, operating under presidential direction. Today, the Bureau of the Budget and its Director are component parts of the Executive Office of the President. Although Congress, in making appropriations, does not always follow the President's budget estimates, it usually comes close to them, and it is difficult to see how an American President today could keep even a semblance of control over the vast federal administrative machinery without his presidential budgetary powers. Many students of American government believe that the President still does not have sufficient control over his administration and that he will not come near to having it until he is given the power to veto individual items in the appropriation bills. The governors of most of the American states now have this vital power, but Congress still withholds it from the President.

As the Coolidge administration recedes into history, it becomes clearer that Coolidge did not assume his role as ameliorator of a budding crisis and use Treasury and Federal Reserve measures to check the speculative boom. And more and more it is being said of Hoover that he did not sufficiently use presidential leadership, presidential powers, and the agencies of government to cushion the effects of the depression and initiate recovery measures.

Franklin D. Roosevelt contributed mightily to the growth and prestige of the presidency. Most of America's leading Presidents have attained fame in history by carrying the nation through one crisis, but Roosevelt carried the nation through two—the Great Depression and World War II. Under him, federal administrative agencies proliferated as never before, enormously increasing the President's administrative responsibilities. Roosevelt prepared a reluctant American public for war;

he exercised immense authority over the wartime economy; he took a personal hand in directing both the diplomacy and grand military strategy of the war; and he was the father of the United Nations. As spokesman for the nation and great teacher, Roosevelt could take the most complex problems and give them a simple clarity, make them understandable to plain people everywhere—and, supremely important, he could do this without robbing them of content and depth, without reducing them to the propaganda stereotypes of "Big Brother." Finally, Roosevelt set up and structurally organized the Executive Office of the President, to make the presidency more manageable.

President Truman had to deal with chaotic world conditions, the collapse of whole societies, which followed World War II. He was called on to exercise powers in the world analogous to those once exercised by a Caesar Augustus, to make specific decisions of the most difficult kind; and he made them. The Truman saga has contributed to the great folk tradition that a person not much gifted above the average run of men can handle the job of President, even the awe-inspiring responsibilities it carries today. However, neither Truman nor Eisenhower had the second Roosevelt's genius for carrying extremely complicated public problems to the people and making those problems understandable. Truman failed to explain the stalemate in Korea, the nature of limited war in the nuclear age. And both Truman and Eisenhower, while including foreign economic aid in their legislative programs, never really succeeded in explaining to the American people the urgent necessity for economic and social programs in the underdeveloped areas.

President Eisenhower, during his first term, did much to relax tensions—the tensions of McCarthyism at home and the tensions of Korea and the Cold War abroad. Under Eisenhower, the Executive Office of the President took its present form. This consists of the White House Office itself, with its numerous top-flight aides and assistant and special aides, and the Bureau of the Budget, the Council of Economic Advisers, the National Security Council, and the Office of Defense Mobilization. Eisenhower sought to simplify the President's huge tasks by operating much like a chief of staff. However, the President is much more than a chief of staff, and he must go far beyond his staff briefings if he is to get an adequate feeling for the many problems confronting him. The Eisenhower administration, by establishing a Cabinet Secretariat and other devices, also sought to arrest the decline of the President's Cabinet in power and prestige.

What forces have been at work during this century to augment the powers of the presidency? Let us summarize these.

First, Americans are now more national-minded, democratic-minded, and mass-minded than they were at the beginning of this century. Increasingly, they have come to regard the President, elected by the whole nation, as more representative of national and democratic opinion than Congress, which they have come to regard as largely representative of local interests.

Second, foreign policy and national defense have now become the major areas of public policy, and since the President must take the initiative and the ultimate responsibility in both these areas, he and his office have loomed larger. Increased power and prestige in these roles have carried over to give him more power and prestige in his other roles.

Third, industrialization, urbanization, and the centralization of the economy have forced the federal government to expand its activities. The Great Depression of the 1930s markedly accelerated a trend already well under way. All of this has resulted in the welfare state and the administrative state. The President now presides over many more departments, bureaus, boards, and commissions, and the federal administrative machinery today is a giant compared with that of the first years of this century.

Fourth, legislation today, dealing as it does with highly complicated matters—involved tax systems, banking and credit, labor-management relations, the commodity and stock exchanges, agricultural price supports, hydroelectric power, atomic energy, and the intricate new armaments and weapons—is more technical, and Congress has been compelled to turn to the experts in administrative agencies for specialized assistance. Increasingly, congressional bills are prepared in the administrative agencies, presided over by the President and staffed by appointees chosen by him directly or indirectly.

Fifth, much of the legislation now considered by Congress, dealing as it does with economic and social matters, touches individuals more frequently and in more aspects of their personal lives than it did formerly. Then too, today people are organized into powerful functional groups, and these groups are most sensitive to the effects of legislation on their members. Senators and representatives must be wary, for if they alienate important groups in their localities and districts they will soon be "lame ducks." On the other hand, the President can afford to take the large view and be "courageous"; his constituency comprises the whole nation and within it are many localities, sections, groups, and interests. If the President alienates some groups and some areas, he can hold onto and even increase his popularity with other groups, other areas, and in the country at large. Hence the President can stand for a comprehensive program of national legislation, with less concern for the adverse reaction of particular sections, groups, and interests.

Sixth, the President has come to represent a more democratic constituency than Congress, and therefore he stands before the country with greater moral authority. Today, both houses of Congress are so constituted that they overrepresent the rural areas and underrepresent the growing populous urban areas. In the Senate, where every state regardless of population has two senators, the sparsely populated and rural states continue to be overrepresented. Even the House of Representatives, designed to represent population, has come to overrepresent the rural areas and to underrepresent the urban ones. The actual drawing of the congressional districts within a state is left to the state legislature, which in most states is flagrantly weighted in favor of the rural areas. In drawing the congressional districts, the rural-dominated legislature favors the rural areas. It is not uncommon for a congressional district in an urban area to represent from 800,000 to 900,000 people while a congressional district in a rural area represents from 200,000 to 300,000 people.

On the other hand, the manner of electing the President gives full weight to the growing populous urban areas. The electoral vote of each state is largely based on population. Moreover, the electors of each state are elected by the whole state at large and not by districts, and hence a vote in an urban area is equal to a vote in a rural area. Thus the President is elected by a more democratic constituency than is Congress, and the growing metropolitan areas, with their labor groups and minority groups, have come to look to the President to defend and fight for their interests.[1]

1. Although since this was written the national House has been reapportioned on a more democratic basis and the metropolitan areas have been better enfranchised, a number of forces still operate, and will continue to do so, to make the President a more effective fighter than Congress for national and democratic causes. Moreover, with respect to the current situation, the reapportionment of the House has not resulted in the increase in the number of liberal members that had been anticipated, for in the meantime rightism has made gains in metropolitan and suburban areas, and this is reflected in the House membership. (This trend to rightism, however, is much less evident in Congress than in a number of the state legislatures, particularly those

Seventh, the presidency has become the beneficiary of the rise of national journalism and the other mass media of communication in the twentieth century. A story about the President, no matter how trivial, is news in the national media because the President is the leading national personality.

The spectacular rise of the presidency during this century should not blind us to the fact that the President, far from being a dictator, actually has more responsibilities than he has powers to meet them.

The President must give leadership to a country of continental size, to the most pluralistic society in the world. A single *state* like New York, Pennsylvania, Michigan, Ohio, Illinois, or California has a wider variety of economic enterprises, occupations, nationalities, races, and religions than any *country* of Western Europe.

The American system of government, like the American society, is highly pluralistic. There is the federal system, which distributes power between the states and the federal government. There is the separation of governmental powers—executive, legislative, and judicial—within the federal government itself. A President may be checked not only by Congress but also by the federal courts, in the final analysis by the United States Supreme Court. America's two major parties, expressing the amazing pluralism and diversity of the American society, are internally divided into many factions, particularly in Congress. The President's own party in Congress is divided, and sometimes a wing of the President's party will combine with the dominant wing of the opposition party to check its own President. About one fourth of the time, too, the President is of one party and the majority in Congress is of the other party. Members of Congress frequently defy their own party, their leaders, and their President in response to interests, groups, and political machines in their own local states and districts.

The President must even bargain with Congress, congressional leaders, chairmen and ranking members of congressional committees, and autonomously situated administrative officials for the control of his own administrative agencies. Congress creates the administrative agencies and prescribes their structural organization. In making appropriations for the agencies, Congress does not always follow the President's recommendations, and through its control of the purse, Congress sometimes

---

in the South and West.) The disadvantaged minorities in the cities still look to the President for their tribune, and when he does not assume that role they feel leaderless.

determines the policies and even the personnel of some of the agencies. There are times when an agency (or a particular administrator) becomes virtually independent of the Executive by virtue of its ability to ingratiate itself with Congress (the Corps of Army Engineers), its close personal relations with a chairman of a congressional committee, its ties with a powerful pressure group (the alliance of the American Federation of Farm Bureaus with certain bureaus of the Department of Agriculture), or its hold on public opinion (the F.B.I. and J. Edgar Hoover). By turning on the white heat of congressional investigation, Congress may make the departments quail.

Not one of the great roles of the President is automatic and self-executing. To have great constitutional powers, to possess strong institutional weapons, and merely to give commands are not enough. A President must never cease being a politician; he must constantly put to use his personal skills and prestige; he must mediate, persuade, negotiate, bargain, and use the carrot and the stick every day he is in the White House.

໕ The Ungenerous Approach
to Woodrow Wilson

Why should an essay on Woodrow Wilson be included in a book which concentrates on the middle decades of our century? Because Wilson's influence has had an enormous impact on all the decades which have come after him, an influence which has been belatedly recognized, and not fully acknowledged to this day.

Even before he entered politics, Wilson was the foremost literary exponent of the "strong" President, and during his presidency he exercised remarkably creative and effective presidential leadership in both domestic and foreign affairs. While the strong presidency was in some ways anticipated by Theodore Roosevelt, it did not come to full flower until Wilson's administration. Franklin Roosevelt followed the Wilsonian example, and by John F. Kennedy's time the active and aggressive presidency had become so acceptable and the old American fear of the Executive had been so overcome that Kennedy, as a candidate for President, found it to his advantage—the first presidential candidate to do so—to promise the voters he would be a bold national leader in the tradition of Wilson and FDR.

Wilson, more than any other statesman of this century, provided the most persuasive rationale for expanding federal activities to solve economic and social problems; his domestic achievements were in many ways a forerunner of FDR's New Deal; and his proposals in world politics, taken together, came to constitute the ideology of the Western democracies, not only in World War I, but also in World War II and in the decades of Cold War which followed World War II.

While much of the following article examines the reasons for the belated recognition of Wilson's high place in history (which in themselves may be a useful commentary on our American political culture), the essay also points out Wilson's substantive impact on the American and the world politics of the decades which followed him.

■ ALTHOUGH THE splenetic atmosphere which enveloped Woodrow Wilson in the 1920s and 1930s, the years of post-Versailles

The Virginia Quarterly Review, XLIV (Spring 1968).

disillusionment, has cleared, a grudging approach to the man survives. If we are to place Wilson in discerning historical perspective, we have still to penetrate persisting myopic misconceptions of him and disclose their ramifying sources.

However, before tackling this contemporary problem of complex and broad historical interpretation, let us take a look at some of the venomous distortions from the earlier period. (The recently published Freud-Bullitt book is a relic of that period.) These constitute in themselves an amazing curiosa, remind us how far we have progressed in better understanding Wilson, and shed light on a continuing unfair approach to him.

William Allen White described young Tommy Wilson as a sickly, bespectacled lad, kept out of the rougher sports by adoring parents, not sent to school and not taught his letters until he was nine, coddled by two older sisters, and "bereft of the inalienable rights of boyhood." Karl A. Menninger portrayed Wilson, throughout life, as schizoid, with a particular "queerness, a queer sort of queerness." William Bayard Hale published a "psycho-semantic" study of Wilson, in which the author, exercising a savage exegesis on Wilson's writings and speeches, with passage after passage torn from context, "proved" that Wilson was a man torn by conflicting emotions, shaken by doubts of himself and his every thought and act, haunted by a never-to-be-extinguished apprehension of his own inferiority. J. M. Keynes, anticipating the Freud-Bullitt devil theory of history, wrote a celebrated volume in which he claimed that the "failure" at Versailles was due to Wilson's personal and psychological flaws.

On the one hand, the liberal Thorstein Veblen saw Wilson's League of Nations as an institutionalizing of *Realpolitik* and of the forces of imperialism, an instrument to preserve the status quo. On the other hand, even so moderate a conservative as William Howard Taft surmised that Wilson sympathized with the Russian Bolshevists. From the left, John Dos Passos heaped merciless ridicule on "Meester Veelson." From the right, H. L. Mencken sneered that "Doctor Wilson" was "the perfect model of a Christian cad." In its first edition, Lincoln Steffens's *Autobiography* displayed photographs of Lenin, Mussolini, and Wilson, all on the same page, under the joint caption: "Three Dictators."

Time has softened the asperities of the post-Versailles years. The coming of World War II, which Wilson had prophesied in the event of the failure of the League of Nations, and the emergence of the United States as the leading international power have served, at least partially, to vindicate Wilson. American historians now usually vote Wilson among the four or five most eminent of American Presidents.

Still, the historical evaluation of the man has not jelled. All too often, in books and articles on Wilson, one detects an ambivalence toward him; a lack of warmth; a nagging and carping approach to him personally and to his achievements; a disparaging tendency to read too much into personal traits which in other historical figures are usually ignored or excused, to distort these traits, and to interpret his public policies in terms of them; a testing of Wilson by higher standards of evaluation than are used in testing other famous men in history; a harping on his inadequacies and disappointments; and a failure to make explicit some of his most notable achievements.

August Heckscher was among the first to point out the persistence of a latter-day ungenerous approach to Wilson. Writing as late as 1958, he observed that Wilson scholars and biographers seem in many cases not to like the man to whom they devote such prodigies of detailed research.

In the early volumes of his monumental life of Wilson, a multi-volume work still in the making, Arthur Link revealed a rather too cool and dispassionate attitude toward his subject, but Link has warmed to Wilson as he has proceeded.

Barbara Tuchman expressed a still not uncommon attitude toward Wilson when in a recent article she referred to him as a man "who habitually evaded reality by refusing to look at it." Like a number of other historians, she also quoted Lindley M. Garrison to the effect that Wilson was the most extraordinary and complex person he had ever encountered, that he (Garrison) never understood him and doubted if anyone else could. The plain inference here is that Wilson was so complex as to be positively "strange."

The youthful Wilson, too, has a way of emerging as an oddity, even at the hands of authors with the best of intentions. Freud and Bullitt were not the first to call attention to the practice of the young Wilson to mount his father's pulpit on weekdays to address imaginary audiences. Even so fair and perceptive a historian as John Blum describes this behavior by the youthful Wilson; but unless the general reader is reminded that most young Americans ambitious to be orators have taken to the woods and the fields to make speeches to the trees and the corn stalks, he is likely to carry away the impression that Wilson was indeed a peculiar lad.

As late as 1956, Alexander and Juliet George, in their study, "Woodrow Wilson and Colonel House," were in search of a psychological clue to Wilson; they suggested that it would be found in Wilson's father fixation, thus approximating the Freud-Bullitt theory.

As to an evaluation of Wilson's public policies, the weight of Richard Hofstadter's interpretations is unfavorable to Wilson. In drawing a rather sharp distinction between Progressivism and the New Deal, Hofstadter necessarily implies that Wilson's domestic program was the mere tail-end of the Progressive era and in no significant way a forerunner of the New Deal. In foreign affairs, Hofstadter infers that Wilson's record was bleak, that Wilson sensed this, and that in the last part of his career he was like a somnambulist moving in a shadowed world, one undergoing self-purification, in search of martyrdom—and more, that the ineffectuality of the United Nations suggests strongly that Wilson and his League of Nations represented an unrealistic approach to world politics.

The most serious criticism of Wilson is that by rejecting the Lodge Reservations, Wilson strangled his own brain child, committed "the supreme infanticide." This is the view of Thomas A. Bailey, and Bailey speaks for a large number of historians and laymen. This interpretation gives too little weight to the reasons why Wilson refused the Reservations; and, more important, it erroneously implies that Wilson's fame in history is tied exclusively to the fate of a single and specific international collective-security organization.

Robert E. Osgood even more emphatically nails Wilson's historical reputation to the League of Nations and Article X. Is this not too narrow a view?

II

Let us first examine the overcritical approach to Wilson's principles and policies, both at home and abroad.

At the time of its enactment, Wilson's domestic program, known as the New Freedom, was generally hailed as a historic breakthrough. During Wilson's campaign for re-election in 1916, despite the later stereotype that "he kept us out of war" was the chief issue of the campaign, Wilson himself and most Democratic campaigners and editors emphasized the domestic achievements. Subsequently, however, the New Freedom came in for much criticism. Some denigrated it as really "old hat," the terminus of the Progressive era. Others argued that it did not advance far enough into fields later brilliantly invaded by the New Deal. Still others claimed that Wilson's program owed more to Theodore Roosevelt than to Wilson himself, that after having castigated TR's New Nationalism during the campaign of 1912, Wilson wound up embracing it.

In many ways Wilson's domestic program was a climax of the earlier Progressivism (in its decentralization of the Federal Reserve system through twelve regional banks, and in its insistence on economic competition as exemplified in the Federal Trade Commission and the business sections of the Clayton anti-trust law); in other ways it was an anticipation of the New Deal (in the labor sections of the Clayton law, in the federal child labor law, in the employers' liability act with respect to the railroads, in the regulation of hours and wages of railway labor under the Adamson law). Many of the Democratic leaders who participated in the battles for the New Freedom also participated in those for the New Deal, and most of these, including FDR himself, emphasized not the differences between these two but their continuity.

During his battle for the New Freedom, Wilson insisted that given the changed conditions of the twentieth century, the extension of federal power for economic and social reform was an enlargement of traditional American liberty and democracy, and not an impairment of them. It was Wilson, more persuasively than TR before him and FDR after him, who gave twentieth-century social reform its intellectual rationale.

Under Wilson the twentieth-century Democratic party became an effective instrument of government and reform. Although historians like Blum and Link give Wilson much credit for this, it is an achievement little recognized by the publicists and the public; indeed, it is one of the most conspicuous instances of failing to give Wilson his due. During his first term, Wilson put together an effective Democratic coalition of agrarian liberals, industrial labor, white-collar people, and intellectuals, a coalition which also attracted many of TR's Bull-Moose followers, a number of Jewish leaders, and a significant minority of businessmen. Even during the 1920s, the years of Democratic defeat, the Democratic party never returned to its Cleveland conservatism or to its Bryan parochialism. Wilson's transformed Democratic party was a clear anticipation of FDR's later winning New Deal coalition.

Wilson is often blamed personally for the violations of civil liberties during World War I, although Lincoln's suspension of the writ of habeas corpus and FDR's concentration camps for civilian Japanese-Americans are usually treated in perspective as defensible war measures.

Because of his repeated criticisms of the balance of power as an ineffectual keeper of the peace, Wilson is often said to have had no under-standing of "power politics." Yet his Peace Without Victory speech is a classic statement of the traditional balance of power, of its shortcomings, and of those conditions under which it may operate with most promise of success. Nor did Wilson ever fool himself about the real reason America entered the war in 1917—to prevent the upsetting of the balance of power by Germany. (However, he publicly put American participation on grounds more congenial to American legal thinking and ideals and more hopeful of better peace-keeping machinery in the future.) Letters, diaries, and memoirs by those close to the Wilson administration have sub-sequently revealed that it was fear of German victory that took America into the war; and Wilson himself bluntly stated, in response to a question of Senator McCumber of North Dakota at a White House conference with senators in 1919, that the United States would have gotten into the war even if Germany had committed no act of war or no act of injustice against our citizens.

Wilson, who fought to save Germany from partition, who prevented France from taking the Left Bank of the Rhine, and who succeeded in scal-ing down reparation claims (all of which were attempts to salvage Ger-many as a power balancer), is said to have had no understanding of the traditional balance of power; but Theodore Roosevelt, who wanted a harsh and mailed-fist peace (which would have left Germany a power vacuum), and FDR, who insisted on unconditional surrender and a dic-tated peace (which indeed left a power vacuum in Central Europe), are held to be "realists" who "understood" the balance of power. (Wilson's Peace Without Victory speech indicates he felt even the League would work better if based on an equilibrium of power.)

Wilson is called "doctrinaire," although he was enough of a pragmatist to fight to save Germany as a power makeweight and to bring back from Paris a proposed American-British-French treaty of alliance to guaran-tee France from further German aggression until such time as the new methods of collective security should become effective. (This treaty was short-circuited in the Senate.)

Even Wilson's use of the Fourteen Points as brilliant war propaganda, which hastened the collapse of the Central Powers, is often conceded reluctantly. As to the substantive merit of the Fourteen Points and other Wilsonian peace aims, critics have claimed that the League was a mere alliance of the victorious powers in defense of their war spoils; that the League, and particularly Article X, froze the international status quo and made no provisions for peaceful changes in it; that too little attention was given to international economic collaboration in an increasingly interdependent world. All of these objections overlook the capacity of new institutions for historical growth, a growth impaired in this case by America's refusal to participate.

More than any of the West's leading statesmen of the time, Wilson realistically understood the stirrings among the colonial peoples and anticipated the importance and nature of the anti-imperialist revolutions (already in the making before World War I), as evidenced by his friendliness to the Sun Yat-sen Revolution in China; his choking off of the Huerta counterrevolution in Mexico, thereby making possible the triumph of the Mexican Revolution; and his mandate system (admittedly inadequate) within the League of Nations. Wilson saw in the oncoming anticolonial revolutions more than mere political nationalism; he saw that these involved basic social revolutions as well. For this historic insight he has been given little credit.

Wilson's most pre-eminent and enduring achievements are still played down or ignored by historians. What are these?

First, Wilson led the United States from isolationism to its first full-scale participation in world affairs, and while there was an interlude of retreat, in the end the Wilson revolution was accepted.

Second, Wilson formulated and articulated the creed of the Western democracies in their twentieth-century struggles for survival. The basic ideology of the Western democracies—during World War I, during World War II, and in the contest with the Communist powers following World War II—has been and still is a Wilsonian ideology: self-determination of nations, anticolonialism, recognition of the social revolutionary implications of anticolonialism, freer trade among the nations, international arms control, international collective security. Essentially, these have been the professed principles of the Western democracies, despite certain obvious deviations from them in fact. Western emphasis and agreement on them have grown since Wilson's time. They serve today as the official long-time objectives of American foreign policy.

Third, Wilson was the first conspicuously placed non-Marxist statesman to escape parochial nationalism and make the breakthrough to internationalism; and he used the power, prestige, and publicity of the American presidency to prepare men's minds for a non-Marxist world order, comprehensive enough, though, to include participation by Marxist powers. (This world order is painfully and tortuously in process of formation today by way of the United Nations and its functional agencies; many functional international organizations, associations, and consortia outside the United Nations; multinational and even supranational collaborations for regional developments; and the growth of a palpable world opinion which includes the neutralists and rises above and cuts across the military alliances and blocs, as discerningly described by J. W. Burton.) This is Wilson's supreme achievement, and compared to this, what particular organizations, agencies, and instrumentalities are developed to implement an international order are strictly secondary.[1]

III

The hypercritical approach to Wilson reaches beyond his policies to his methods in politics, his techniques, his style.

Even Wilson's remarkably effective use of the English language is held against him. The word "rhetorical," with its suggestion of artificiality and showiness, is used over and over again to describe Wilson's style. (It is strange how rarely that term is applied to Kennedy's inaugural address, which was indeed a contrived bit of rhetoric.) Wilson's speeches are eloquent blends of thought and feeling, lucidity and grace; his oratory, even more than Churchill's, moved men's minds, hearts, and hopes. Compared to Wilson, Bryan is spread-eagle, Lloyd George is ephemeral, FDR is journalistic, Stevenson is precious, Kennedy is declaratory, and Churchill is declamatory and, to quote Malcolm Muggeridge, sometimes even "gaseous." But the Wilson critics, even in stylistic matters, are insatiable. Professor Bailey rebukes Wilson for falling into unbecoming colloquial

1. The unique significance of this is more readily realized when Wilson's approach is compared to that of his American contemporaries, among them John Hay, Elihu Root, Theodore Roosevelt, William H. Taft, Philander C. Knox, Henry Cabot Lodge, David Jayne Hill, and William Jennings Bryan, who "tiptoed" to "internationalism" by way of legal arbitration of disputes between nations. These argued endlessly about arbitration and conciliation, about what questions were "justiciable" and what were not, about what should be subjected to compulsory arbitration and what left to voluntary arbitration. Although Wilson's League of Nations provided for the judicial settlement of disputes, he saw that the questions which arouse the passions of nations and lead to international crises and war are in the main not legal but political, and that only a political internationalism could deal with the economic, social, and political dynamism which matters most in world politics. Wilson did not believe that this kind of internationalism would develop overnight; his League of Nations was only a beginning foundation upon which, hopefully, an evolving functional international politics might build. Technological forces and a palpably developing world opinion may ultimately converge to raise the United Nations and other international agencies from a secondary to a primary place in world politics.

expressions during his 1919 speaking tour on behalf of the League. Wilson, however, in his political campaigns, had always used colloquialisms with telling effect, and these speeches of 1919 deeply moved their audiences. They were marvels of eloquent extemporaneous speech-making, for at every stop essentially the same substance was repeated in different language and imagery.

Since it is manifestly difficult to dismiss the palpable impact of Wilson's oratory throughout his career, Wilson comes in for much more criticism on the score that he was deficient in the world of action, particularly in his relations with individual politicians. Actually, Wilson had a far easier give-and-take with the politicians than is usually supposed. Wilson liked banter, a good story, or a tart rejoinder. And he could laugh at himself, even at his Calvinism. For instance, Marquis James recounts how John Nance Garner visited Wilson one night after a particularly grueling presidential day. Wilson told Garner that if it were not for his faith in the old Presbyterian belief in predestination he didn't think he could carry the presidential burden much longer. Garner replied that he understood, that "having grown up in the Red River Valley with all those darkies" he, too, was "superstitious as hell." For a moment Wilson hesitated, then he leaned back and laughed heartily. "Garner," said the President, "I think our superstitions will carry us both through." Incidents like this, which reveal the real Woodrow Wilson at work with the politicians, are legion; but historians who confine themselves to the official source-collections miss them, for they must be gleaned from the diaries, letters, and biographies of the run-of-the-mill politicians.

It is difficult to gainsay Wilson's remarkable success in dealing with Congress as long as the Democrats were in the majority there. This success was due to the definiteness and clarity with which Wilson formulated his legislative programs; his dramatizing those programs by personally addressing Congress in support of them; his many personal conferences with legislative leaders and rank-and-file congressmen; his personal mediation between the various groups which composed his party coalition; his willingness to make the compromises inherently necessary in leading a party combination so wide-ranging and diverse; his appeals to party loyalty and use of party patronage to keep even the Southern conservatives in line; and his direct appeals to the country at large. Wilson was severely criticized for calling for the election of a Democratic Congress in 1918, when the war was approaching its end, and some historians still criticize him for this. But that appeal was consistent with Wilson's view of strong party government; he felt that the Democrats were better friends of his peace aims than were the Republicans; and since Wilson's time most American Presidents, Republican and Democratic, as a matter of course have asked for the election of a Congress of their own party. Wilson gave party and presidential leadership a larger dimension, and our contemporary conception of the "strong President" owes much to him.[2]

Not only did Wilson get along well with Congress during most of his tenure, but his relations with his administrative subordinates were also remarkably harmonious. Wilson was an efficient administrator, among the most efficient of all the Presidents; indeed, he was the ablest war administrator in our history. The overlapping of functions, the clash of personalities, the in-fighting, the playing of one agency off against another—all so conspicuous under FDR—were kept at a minimum under Wilson. The corruptions and scandals of wartime, much in evidence in most of our wars, even in our little war with Spain, were relatively absent under Wilson. Also lacking were the fierce political rivalries among the generals and the playing of politics in the appointments of commanders, which have given most of our wars both tragic and comic aspects. In 1917, America was faced for the first time with having to wage the total war demanded by modern technology, and the Wilson administration hastily assembled a galaxy of war agencies which did the job, agencies which later became points of departure and even models for some of FDR's New Deal and World War II agencies. Here again, in our infrequent acknowledgment of Wilson's great ability as an administrator, is another conspicuous example of the reluctance to give Wilson his just deserts.

Frequently more attention is given to Wilson's breaks with some of his personal advisers and political lieutenants than to his flexible, harmonious, and sustained relations with Congress, his party, and his administrators. There were a number of personal breaks—among them, those with George Harvey, William F. McCombs, E. M. House, Robert Lansing, and

2. Despite his belief in an aggressive presidential leadership, Wilson scrupulously respected the powers of Congress. As a wartime President, Lincoln went ahead and exercised extraordinary powers and then later asked Congress to ratify them. The source of Lincoln's power was a broad construction of presidential war powers under the Constitution, and he operated in spite of Congress. On the other hand, Wilson as wartime President worked in harmony with Congress. Outside the area of military command, Wilson exercised extraordinary war powers only after they had been specifically granted him by Congress.

Joseph Tumulty. Some of these ruptures have an admittedly testy quality. But is there not an overemphasis on them? Because of its very nature, any political career will reveal numerous shifts in personal relations and political alignments. Lincoln never hesitated to throw overboard a subordinate, colleague, or political ally when it appeared necessary to his career or the public interest; his abrupt dropping of Montgomery Blair from his Cabinet is only one instance of this. Surely Wilson's breaks were not as rancorous, shattering, or history-making as the more numerous quarrels and feuds of Andrew Jackson. Historians do not accuse Jackson of being "neurotic" because of his venomous vendettas, and no psychiatrist has yet suggested that Jackson thought of himself as a Jesus Christ with a need for a Judas Iscariot. (How enraptured the Whigs would have been with *that* analysis!) Again, is not a more exacting yardstick being used to measure Wilson than is used to measure other historical characters?

This harping on Wilson's personal breaks ignores his other and far more numerous relationships which never ended in rupture. In fact, the number of personal and political allies who cleaved to Wilson to the very end was exceptionally high—among them, to mention only a few, were Franklin Roosevelt, Josephus Daniels, William G. McAdoo, Newton D. Baker, Carter Glass, John Sharp Williams, Samuel Gompers, Bernard Baruch, Herbert Hoover, Louis D. Brandeis.

What was noteworthy about Wilson was the depth of respect and admiration for him among his intimates, advisers, and political lieutenants. Indeed, there is something of adoration in the respect Wilson's coworkers accorded him; most of them carried this adoration of him to their dying days; even Tumulty never ceased to venerate "the Governor." Franklin Roosevelt often spoke warmly of his attachment to Wilson.

Did Wilson allow personal "vanity" to becloud his conduct of public affairs? Did he go in person to the Paris Conference because of a "messianic complex"? This picayune charge arose in 1918 because Americans of that time were not habituated to playing a leading role in world politics, to having their Presidents attend international conferences on foreign soil. Since World War II it has become commonplace for American Presidents to travel abroad and attend international parleys. Did Wilson operate at Paris on his own, without expert advice? On the contrary, he took with him to Paris the largest battery of specialists and men of large practical experience in world affairs ever assembled by an American President at an international conference. However, the advice of the experts did not always prevail, for the swirling political pressures and counterpressures could not be ignored, and Wilson recognized the need for give-and-take in diplomatic negotiations as well as in domestic law-making.

The image of Wilson as a noncompromiser, an intransigent, rests largely on his refusal to accept the Lodge Reservations to the Versailles Treaty, admittedly the most controversial act of his career. These Reservations required that before United States adherence to the Treaty would become binding, every signatory nation would have to acquiesce in all of them. In a brilliantly researched book, *The United States and the League of Nations*, D. F. Fleming has shown that acceptance of all the Reservations by all the other nations was by no means a certainty, that the United States might well have been put in the position of being rejected, a position which would have still further intensified nationalism in the United States and tensions abroad. Again, had Wilson shown signs of going along with the Reservations, would not Lodge then have raised the ante? Had not Lodge aggressively "crowded" Wilson after every Wilson concession? Would not a further yielding by Wilson have resulted in even stiffer reservations being added? Was not Lodge using the Reservations to achieve not ratification but rejection, knowing that there was a point beyond which Wilson could not yield? In the spring of 1920, neither Wilson nor the Republicans could foresee the Republican landslide which developed in the fall of that year, and hence Wilson's belief in the spring that he might get a better deal from the people than from the Republican Senate did not appear to be groundless. The motives of Lodge, however, no longer admit of doubt, for following the Republican triumph in the election of 1920, which gave the Republicans the presidency and a large majority in a Senate dominated by Lodge in foreign policy, the victors threw overboard their own Reservations and turned their backs on the whole Treaty, including the League.

IV

The double standard of evaluation applied to Wilson reaches to even his private and personal traits.

Many writers have turned the young Tommy Wilson into a scarecrow, portrayed him as physically unattractive, sickly, sissified. Wilson's intimate contemporaries, those who had been his boyhood and youthful companions in Augusta, Wilmington, and Columbia, were revolted by this grotesque caricature; it bore no resemblance to the Wilson they knew. Among Wilson's early companions, there is a remarkable consensus about his appearance and character as a boy and youth. They saw nothing unusual in Wilson's appearance, health, or hibits. A few sensed in him his mother's reserve, but most remembered his sociability, his taking a lead in activities, and his smile, which was infectious when he flashed it. "Stocky" was the one word used over and again to describe Wilson's early appearance. Pleasant A. Stovall, a boyhood chum in Augusta and in mature life publisher of the Savannah *Press*, recalled that Wilson was an inventor of pranks, a member of the gang's baseball club, and a devotee of riding horseback. John D. Bellamy of Wilmington recounted how he and Tommy spent hours down at the port inspecting foreign ships and dreaming of going to sea and how Wilson would propound knotty military problems for his young brother Joe's gang to work out in their war games. Later, Wilson played baseball at Davidson, was elected president of the baseball club at Princeton, invented deceptive plays as part-time football coach at Wesleyan, tossed the most professional ball of all the Presidents at the opening of the Washington season, and learned to play a fair game of golf.[3]

In any event, Wilson's boyhood and adolescence were not nearly as much marred by illness as were Theodore Roosevelt's; and Wilson was a healthier man during most of his presidency than FDR and Kennedy were during their whole tenures. Yet Wilson's "sickliness" is played up, and the ill health of TR, FDR, and Kennedy played down.

Did Wilson display weakness in depending financially on his father until his late twenties? What is unusual about persons of intellectual bent prolonging their maturing process, probing and exploring widely before they find themselves? The fact is that Wilson did finally become financially independent, but when did Theodore Roosevelt, Franklin Roosevelt, or John Kennedy ever cut off their financial dependence on family wealth?

Even Wilson's pre-political career is often belittled. Critics claim that as a teacher Wilson was primarily a "performer" (a charge often made by academicians against their more popular colleagues), and that as a writer on politics he was a "mere" generalizer. Nevertheless, Wilson's pre-political career as teacher, educator, writer, and brilliant interpreter of political institutions was a distinguished one. On the other hand, it is sel-

---

3. Some have expressed disbelief that Wilson, at any stage of his physical growth, would have been called "stocky." The records of Pleasant A. Stovall and other boyhood and youthful companions are clear on this point, and a glance at photographs of Tommy Wilson taken with classmates at Davidson, Princeton, and the University of Virginia bear them out, for the broad and square shoulders of Wilson make him stand out from the crowd.

dom held against TR, FDR, and John Kennedy that they had no regular or sustaining pre-political profession of any kind, that all of them had been dilettantes and dabblers.

Was Wilson a prig and a pedant, a man devoid of the social graces? On the contrary, he mixed more easily with the politicians than is commonly supposed, and he was a delightful social companion. Many who spent an evening with him and his family in the Princeton home or in the White House have left accounts of the light-hearted atmosphere of these occasions. Wilson had a large fund of human-interest stories, liked to read poetry aloud and recite limericks, and had a happy gift of mimicry. Wilson was attractive to women; both his first and second wives adored him, and he was devoted to them. He derived stimulation from platonic friendships with vivacious women, from the kind of conversational companionship Henry Clay found in Harriet Martineau; and he was an inveterate theater-goer.

V

Why the lack of warmth toward Wilson? Why the ambivalent and even grudging approach to him?[4]

The climate of opinion at the time a conspicuous figure leaves the stage of history has always influenced his historical reputation, particularly during the first succeeding decades. It is clear now that American opinion in 1920 was sick and, toward Wilson, rancorous. There were a number of reasons for this. Among the more important were the pains suffered by an advanced industrial society in converting from a war to a peace economy; the Red scare and witch hunt produced by hysterical fear of Bolshevism; the angry emotionalism of German-Americans, Irish-Americans, Italian-Americans, and other minority groups alienated by Wilson's foreign policy. Other causes, however, were more pervasive. There was a delayed but intense resentment over the unprecedented regimentation required by modern total war; in the immediate let-down from war the omnivorous wartime controls appeared, in retrospect, to have been in-

4. In discussing the reasons for the delay in recognizing Wilson's historical achievements, the author here makes no mention of the partisan rancor at the close of his administration, a rancor comparable to that of Jackson's day and of the Reconstruction period. Why the Republican animosity to Wilson reached such a pitch of intensity and the part it played in postponing an objective historical evaluation of him should have been included. The main reason it was not is that this aspect of the subject was discussed by the author in an earlier article, "A New Look at Woodrow Wilson," published in the *Virginia Quarterly Review*, Autumn 1962.

tolerable. Again, the American plunge into full-scale world politics in 1917, although necessary to defeat Germany, was bound to recoil, for it was too wide and abrupt a break with the American past. Such a plunge would have been difficult to assimilate in any event, but it took place at a time when international relations were extraordinarily turbulent. Fierce forces of national, anticolonial, and social revolution were unleashed by the war, and the traditional balance of power was disintegrating. Wilson had anticipated these developments, and his Fourteen Points and other peace aims were attempts to accommodate them. But no American President could have prepared the American mind for total and permanent immersion in world politics on such short notice. Even in 1945 that immersion would be difficult to absorb; in 1919 it was impossible.

There are, however, more enduring reasons for the hypercritical approach to Wilson.

Wilson had suffered from the nature of his pre-presidential career. Americans were accustomed to picking their Presidents from lawyers, planters, generals, and career politicians. A college teacher in politics, above all in the presidency, was a curiosity. Even the British would have been a little patronizing toward a professor in the premiership. The academician in politics is a continental tradition, not an Anglo-American one. During Wilson's presidency, and since that time, Americans have been over-prone to see in Wilson the stereotyped traits they have associated with academicians: aloofness, preciousness, doctrinaire attitudes, an over-respect for theory, an avoidance of reality.

It is true that Wilson was less extroverted and more introspective than most American politicians, and he never simulated a folksy heartiness. Jefferson, Lincoln, McKinley, and FDR were also introspective, and in addition they were secretive—and in all candor a bit devious. But these men had protections which Wilson lacked. All had been long-time political practitioners at the grass roots. Jefferson was favored by his Revolutionary fame, his membership in an acknowledged ruling class, his legion of friendships with local politicians cultivated over the years. Lincoln was shielded by his humble origin (consciously inflated into a popular myth during his lifetime), his *appearance* of folksiness, and his deliberate practice of diverting attention from his inner thoughts by telling one salty anecdote after another. McKinley encouraged stories about his piety, his love of old hymns, his solicitous care of his invalid wife. FDR was ambidextrous and the most nimble; he could move back and forth from extroversion to introversion, from introversion to extroversion.

Wilson himself was in large measure responsible for the general public's lack of familiarity with his courteous, warm, and genial traits in private life and personal intercourse. Although Wilson institutionalized the presidential press conference, he drew a sharp distinction between his public life and his private life. He believed that what he did as a public man was news, but that what he and his family did in private was not news. The American press and public insist that *anything* a President and his family do is news; but Wilson emphatically rejected this. Most American Presidents have encouraged human-interest stories about goings-on in the White House, but Wilson positively discouraged these.

In a sense, too, Wilson was the victim of his own efficiency, for had there been more slambang fights within and between his administrative agencies he might have come off as a more colorful and "human" individual; and the press would have had the journalistic field days it adores.

In another way, too, Wilson contributed to the tendency to judge him overcritically, to measure him by higher standards than are used to measure others. Wilson often saw more clearly than his contemporaries the problems of his time. However, he had a way of verbalizing his proposed solutions and his aims in idealistic terms. When his goals were essentially realized, as they frequently were, they sometimes fell short of the expectations he had aroused. But had he not aroused those expectations, he might not have achieved the goals. Many have judged Wilson not by his actual achievements but by the visions he raised to win them; they have measured him not by the ordinary standards of history but by his own loftier ones.

Only twelve years after Wilson left office, Franklin Roosevelt came to the presidency and, like Wilson, enacted first a reform program and then led the nation in world war. Since FDR's reform program was wider and American participation in World War II longer and more extensive, FDR, in the short run, came to overshadow and dwarf Wilson. It will take a longer time for us to see clearly that it was Wilson who was the true trail blazer, the one who broke ground for FDR's New Deal, and, infinitely more difficult, for America's full-scale participation in world affairs.

Wilson has become the tragic figure in one of world history's most gripping dramas. Rather than the statesman who succeeded in achieving many notable things, he has become the prophet who failed. In concentrating so exclusively on the miscarriage of the League of Nations, we have obscured other aspects of Wilson's career. All too often the story of Wilson has been told with the failure of the League always in view; we are constantly looking for hints and suggestions during the whole of his life that will help explain a tragic climax in personal terms. Not until we give less attention to Wilson the martyr and more attention to Wilson the statesman will we put him in accurate historical perspective and discern clearly his positive achievements in domestic and world politics.

Because, among other things, of the tendency to emphasize Wilson's career as climaxing in personal tragedy, his life has lent itself to the psychoanalytical approach, that is, to the earlier and cruder Freudianism, now outdated and undergoing modification, but unfortunately still prevalent. This early Freudianism reduces the ego to a mere battleground for the war between the id and the superego, conceived of as deterministic products of either biological or social conditioning, and when applied to historical figures it tends to convert them into case studies and patients, to "trivialize" and "puppetize" them. Those who fare worst are the "troubled" personalities—the reformers and crusaders, the movers and shakers, the seers and prophets—who are usually made to appear ridiculous and contemptible. It amounts to discrimination to single out a few historical characters for this crude kind of Freudian treatment and leave others untouched by it.

If this approach were to continue long enough, however, more and more historical leaders would be subjected to it, and there would be less biographical discrimination; but then history, largely reduced to a succession of predetermined behavioral activities, would itself become a grand distortion. We would perhaps discover that Lincoln's history-making rejection of the Crittenden Compromise was compelled by Lincoln's inner "complexes" (growing out of his sensitivity to his physical awkwardness, his bleak childhood and youth, his lack of formal education), which drove him to "prove himself" in this first major test of his leadership; that Theodore Roosevelt's combative "neurosis" (born of a need to compensate for early ill health, defective eyesight, and a high-pitched voice) impelled him to disrupt the Republican party in 1912, thus giving Wilson's "neuroses" their opportunity to make history.

Fortunately, ego-psychology itself is moving away from its early crude Freudian determinism and redirecting its attention to the self-strivings of men, with an appreciation of the impact of life's repeated emotional experiences on the growth of personality. As Cushing Strout has pointed out, there is now a new promise, still difficult to fulfill, more difficult than advocates of interdisciplinary scholarship usually suggest, of combining a more mature psychoanalysis and a more responsible history "in order to focus on those complex intersections of men and movements when reason and the nonrational somehow come together." Under this kind of psychological-historical treatment, we may be sure that the long-time historical reputation of Wilson would not suffer.

However, another development may postpone a full realization of Wilson's great stature in history. Philip Rieff has observed that there is now taking place a general assault on all inherited political styles, including the Communist. "Lenin would have done just as well as Wilson, though Gladstone makes a more familiar candidate." Rieff suggests that a new cultural type may be emerging—the anti-leader.

The charismatic leader, with his faith in large ideas, insights, and principles, is currently at a discount. Critics of the old-style politics, converging from many disciplines, fancy they are imposing a new style, in which political leaders largely become facades for technicians, who operate as collegial task forces to solve our public problems "scientifically." Now, in its first flush of success, every historical movement carries a fair share of utopian baggage. The new technocratic society has its own illusions and superstitions—that nontechnocratic values are on the way out, that social and group conflicts will yield to expertise, that the complexities of history can be left to the neat disposals of the bureaucrats. A true appraisal of Woodrow Wilson's stature in history may have to wait on some receding from this current intellectual climate.

❧ The Revolution in the Presidential Nominating Convention

More than a decade has passed since the following article was published. In the light of subsequent events, it seems to have captured the long-time trend to democratizing our method of making presidential nominations but to have exaggerated its pace.

Written before the presidential election of 1960, the article clearly anticipated the ways the nominations would be made that year. Nixon, the obvious national favorite among Republicans, was easily nominated. Kennedy, who by 1960 had deserted the congressional cabal which supported him for the vice-presidential nomination in 1956, successfully used all the modern methods of mass democracy—Madison Avenue techniques, mass media, opinion polls, and popular primaries—to win the Democratic nomination hands down over Senator Lyndon B. Johnson, who relied on the old-fashioned ways.

The elections of 1964 and 1968 were less clear-cut examples of the trend. In 1964, sitting President Johnson was of course nominated by the Democrats. Barry Goldwater, the Republican nominee, got the most dependable part of his delegate strength from those states which had selected their delegates in state conventions and not primaries, but it is doubtful that Goldwater could have been nominated had he not won the decisive California primary over Governor Nelson Rockefeller. In 1968, Nixon, who won handily all the primaries he entered, was the national favorite among the Republicans, and he was nominated. But Hubert Humphrey, the nominee of the Democrats, owed his nomination largely to the old-time methods and the support of the professional politicians. This throwback, however, should not be exaggerated, for the candidates winning the primaries and making their appeals to the rank and file—Senators Eugene McCarthy and Robert F. Kennedy—were robbed of their chance to unite their forces against Humphrey. The violent death of Senator Kennedy left his supporters dazed, leaderless, and without heart or capacity to make an effective coalition with anybody. Had Kennedy lived, he might have emerged as the unbeatable national favorite by the time the convention convened. As it turned out, by the time the convention met Humphrey

Political Science Quarterly, LXXII (June 1957).

had the support not only of the professionals but also of the Democratic public opinion polls.

The trend analyzed in the following article is likely to continue and suffer fewer exceptions with the passing of time. The Democratic convention of 1968 democratized the future convention rules, which should make it easier hereafter for the national favorite to win nomination. Should the trend to nominating the national favorite fail to develop fully, we are likely to see the finish of the conventions and the establishment of a presidential primary, which will include the contests within all the parties on a single national ballot.

Within the next decade we shall probably abolish the electoral college and adopt the direct popular election of the President. This will tend to multiply parties, and not infrequently it may happen that no nominee receive the forty percent of the popular vote that will almost certainly be required to elect, necessitating a runoff election. If in addition we had a national presidential nominating primary, and if in some of the parties no nomination was made in the first primary for lack of any candidate's receiving the required number of votes to win nomination (for even a national favorite might not win against a field of half a dozen or so other candidates in a first primary), then a runoff would be required. Every fourth year, then, we might have *four* presidential campaigns—a first primary, a runoff primary, a first election, a runoff election. Then the cry would almost surely go up: "Let's prolong the President's term to six years, so we won't have to go through this election turmoil every fourth year but only every sixth year." An agitation for a single, six-year presidential term is already well under way, with Senator Mike Mansfield in the lead.

In other countries there are conventional ways of getting rid of presidents and prime ministers before their normal terms have expired, but in the United States the President's fixed term of years is confirmed by both law and custom. No President has ever resigned, declared himself incapacitated, or been removed on impeachment charges; and Americans pride themselves that no President has ever been "eased" from office by conspiracy, cabal, coup, or *Putsch*. This has worked as well as it has because of the relative shortness of the presidential term. But what if Americans were required to suffer an ineffective or a disastrous President for six years rather than for four, with no acceptable way of getting rid of him short of that time? Would not this invite explosive situations? What would have happened if Americans had been required to keep Pierce another two years? Or Buchanan? Or Andrew Johnson? Or Hoover? Had Lyndon Johnson's term run to 1971, how could he have made his timely and graceful exit?

America's presidential politics have already reached a fast pace and high pitch unmatched in the politics of any other country. Direct election of the President would further prolong and intensify electioneering, and a national primary still more. In the light of all this, would it not be better to encourage the trend to democratizing the conventions, as described in this article, rather than adopt a national primary and thereby add to the frenzy of the presidential election every four years and increase the risk of the nation's ultimately resorting, in exasperation, to a six-year presidency as a means of reducing the frequency of the presidential election ordeal?

■ DURING THE PAST three decades, the presidential nominating convention has been undergoing a major transformation. Party methods of nominating presidential and vice-presidential candidates are yielding, without much formal or structural change, to twentieth-century mass democracy. Delegates to national conventions, even the biggest of the "big shots," are in the process of being reduced to popular rubber stamps, very much as presidential electors were reduced to nullities during the first decade of our present federal Constitution. Increasingly a national nominating convention is merely choosing its nominee from among popular national favorites; increasingly it is being forced to pick *the* national favorite. The days of the favorite son, the dark horse, the stalking horse, the smoke-filled conference room, the senatorial and congressional cabal, and the decisive trading of votes by local bigwigs are numbered, if indeed they are not already finished. "Insiders" and members of the political "club" are being cut down to size and forced to accept the leadership of those who have made successful national preconvention campaigns and of those who have become national "names" and mass celebrities.

More time will elapse before this trend is commonly recognized. Even when it is more firmly established than it is, the temptation of the journalists and magazine writers to play up the color and suspense of the old-fashioned convention, to anticipate the maneuvers of the behind-the-scenes manipulators and the alleged king-makers, of the Thurlow Weeds, the Mark Hannas, and the Jim Farleys, will be strong. Undoubtedly the prospective strategic and "decisive" moves of the Lyndon Johnsons, the Frank Lausches, the Stuart Symingtons, and the Carmine De Sapios of the day (which in fact will *not* materialize) will be widely trumpeted to the public. Especially will this be true of the next Democratic National Convention, for Democratic leadership during the next four years will be in Congress, and Democratic congressional leaders will be blown up to balloon size, only to be rudely deflated at the next convention.

The sooner we realize the altered nature of the nominating convention the better off we shall be, for some of the problems already produced by the trend of the past three decades demand thoughtful consideration now.

II

In 1956, the conventions of both parties again strongly verified the trend of the past thirty years. Eisenhower, of course, was the undeniable national favorite of the Republicans. By way of the long, grueling primary route, Stevenson, by the time of the Democratic convention, had emerged as the undoubted national favorite of the Democrats. Moreover, the conventions of 1956 revealed that the trend is also becoming applicable to the vice-presidential nominees. Even had Harold Stassen begun his "Stop Nixon" campaign six months earlier and enlisted the support of many of the most powerful of the politicians of the international wing of the Republican party, it is doubtful that Nixon, who had become a national "name," could have been stopped, short of direct and vigorous intervention by President Eisenhower himself. But the most telling illustration of the trend to nominate only national "celebrities" is afforded by the dramatic convention conflict over the Democratic vice-presidential nomination.

There is no doubt that Estes Kefauver, by the time of the opening of the Democratic National Convention of 1956, had become a national "name." His television performances as senatorial "crime-buster," his national run for the presidential nomination in 1952, and his arduous primary fights in all parts of the country in 1956 had made his name and personality well known to the country. At the Chicago convention of 1956 he was forced to organize a sudden and spontaneous drive for the vice-presidential nomination against the forces which in the past have been "king-makers" in national conventions. Kefauver faced the bitterest kind of opposition from these traditionally powerful forces. He was opposed by the big-city bosses, who remembered all too well his exposure of the alliance of the underworld and the city politicos. He was opposed by the Southern delegations, who usually follow local and personal leadership to a greater extent than do delegations outside the South and who regarded Kefauver as too liberal and as a renegade to white Southern sentiments on racial questions. He was overwhelmingly opposed by congressional Democrats, by congressional cronies and "insiders," who felt that Kefauver did not belong to their "club." Any acute observer who watched the convention proceedings soon became aware of a congressional cabal, more particularly of a senatorial cabal, bent on defeating Kefauver. The maneuverings of Senators Kennedy, Gore, Johnson, Smathers, Fulbright, Monroney, Symington, Long, and even Humphrey, and of Speaker Rayburn, Representative McCormack, and "Les" Biffle were quite obvious.

Now it is important to note that even powerful members of the Congress are not, by today's standards, national "names." They may head important committees, now and then get their names in the headlines of the inside pages of the New York *Times*, and occasionally "make" the national press associations; but today all of this is a far, far cry from becoming a familiar name and personality among the mass of people. To become a national celebrity, it takes more, too, than a human-interest story in a national mass-circulation magazine, a story predicting how such and such a governor or senator is maneuvering himself into a position of "king-maker" in a forthcoming old-fashioned national convention. All of this seems to come as a surprise to congressional leaders, who are accustomed to being very big men in congressional chambers and lobbies, and who collectively, as the Congress, can on occasion still be very potent indeed. But increasingly presidential politics involve popular group and mass activities in the country at large. Presidential politics are being divorced from local and congressional politics. Hence the growing separation of presidential Republicans from congressional Republicans, of presidential Democrats from congressional Democrats. When congressional leaders at national conventions throw bouquets at one another, as they do so profusely, they do not seem to be aware that the country does not know and appreciate them in the way they know and appreciate one another.

Powerful, then, as the city bosses, the Southern leaders, and the congressional bigwigs would have been in a Democratic national convention of the nineteenth century and of the early twentieth century, they were no match, even in a *vice-presidential* contest, when pitted against a national "name" with a grass-roots national following and the backing of national farm and labor organizations. Undoubtedly, too, Senator Kennedy, the convention runner-up against Kefauver, did better than otherwise he would have because of his Catholicism. Many a New Deal delegate (particularly from Massachusetts and New York), many a delegate who ordinarily looks at politics from the national or presidential angle, deserted his normal political allegiances to seize this unusual opportunity to vote for a fellow Catholic. However, the failure to nominate for vice-president Kefauver, the runner-up in the popular primaries, would have been followed by a keen sense of frustration by rank-and-file Democrats the country over.

What is most important about the results of the Democratic and Republican conventions of 1956 is that they confirmed a presidential nominating trend which has been well in evidence in all presidential nominating conventions following 1924. For after 1924 the convention of each party has picked a national favorite—usually *the* national favorite—as the presidential nominee. The last time a Republican nominating convention chose a favorite son (after national favorites had deadlocked the convention) was in 1920, when Harding was nominated. The last time a Democratic nominating convention chose a favorite son (after national favorites had deadlocked the convention) was in 1924, when John W. Davis was nominated.

III

The trend of the past thirty years is unmistakable. Let us trace it.

In 1928, Herbert Hoover, opposed for the Republican presidential nomination by a miscellany of unimpressive favorite sons, was the overwhelming choice of the rank-and-file Republicans of the nation. He was easily nominated. Likewise, for months prior to their Houston convention, the nomination of Al Smith by the Democrats was a foregone conclusion. Indeed, the choice of Houston, a Southern and Protestant city, as the place for holding the Democratic convention was a precalculated attempt to make Smith's almost certain nomination palatable to Southerners and Protestants. Smith was clearly the logical candidate. In a strongly Republican era, he had been elected governor of New York four times, when other Northern Democrats had gone down like ninepins. Two of Smith's triumphant elections had taken place since the Madison Square Garden convention of 1924. Under these conditions, most Democrats throughout the country felt that to turn down Smith would be a rank injustice and a positive insult to the Irish Catholics, who long had been a bulwark of Democratic strength, and to other "nonnativist" Americans of the large cities.

In 1932, Hoover was obviously the choice of the Republicans; and Franklin D. Roosevelt, the popular governor of New York, was the outstanding national favorite among the Democrats, as evidenced by primaries, national opinion polls, local straw votes, and other barometers of public opinion.

In 1936, Roosevelt was the inevitable Democratic nominee. By the time the Republican convention opened, Governor Alf Landon, one of the few Republican survivors in high office, was far in the lead for the Republican nomination.

In 1940, mass pressures had been built up for Roosevelt's third nomination, and those parochial forces formerly so important in national conventions (in this case represented by old-fashioned politicos like "Jim" Farley and "Jack" Garner) were powerless to stay the mass trend. The Republican convention of 1940 revealed even more dramatically than the Democratic convention the enormous strength of the new mass forces over the old parochial ones.

Mass pressures were responsible for the nomination of Wendell Willkie. The average Republican delegate did not want Willkie. The average Republican politician, guided by the old rules of the game, had every reason to be suspicious of him. The old rules demanded a professional, a proved vote-getter, and Willkie was a rank outsider, an amateur who had never tested his mettle with the voters. Moreover, he was a big-business man, a Wall Street businessman, a businessman identified with the most unpopular of big business, public utilities. And, most unholy of unholies, Willkie was not even a good Republican, if indeed he was a Republican at all. He had been a lifelong Democrat; only as late as 1938 had he formally renounced the Democrats and become a Republican. The nomination of Willkie defied all of the old rules; he was nominated solely because of mass pressures and the historical logic of the situation. (The historical logic of the situation is looming larger and larger in all of our political decisions as we attempt to wrestle with the increasingly remote, impersonal, and global problems of our time, and it will bulk even larger in the future.) It so happened that Willkie's three chief contenders for the nomination—Taft, Dewey, and Vandenberg—represented more or less isolationist records. Willkie had no isolationist record, and he tended to be an "interventionist." Willkie's position, in contrast to that of his opponents, had by the time of the Republican convention become a decisive advantage, for France had fallen and the whole country was watching the battle of Britain with breathless anxiety. Americans were frightened, and American opinion with respect to the foreign situation was moving rapidly away from isolationism. Moreover, many of the nation's magazines of mass circulation, most conspicuously the Luce publications, had been sounding a steady drumfire for Willkie for weeks before the convention opened. His nomination, then, represents a striking victory for mass historical forces, the techniques of Madison Avenue advertising experts, and the agencies of mass opinion—polls, radio, and big-time journalism—over the traditional, leisurely paced rules of American politics.

In 1944, the logic of the historical situation made Roosevelt, America's wartime leader and world spokesman, the Democratic nominee. The national favorite among the Republicans was Thomas E. Dewey, who had progressively modified his old isolationism and who, because he was governor of New York, the most populous state in the Union and a microcosm of the nation's racial, religious, national, and economic groups which bulk so large in contemporary national welfare politics, was by all odds the most "available" Republican in high elective office.

In 1948, Truman, a sitting President and world leader in time of crisis, was in a position to dictate his own nomination. Dewey, in spite of his defeat in 1944, was the logical Republican choice. Since that time he had been re-elected governor of New York, and in a series of hard-fought primary contests with Stassen and others, he had emerged as the undoubted national favorite among the Republicans.

In 1952, the victory of Eisenhower over Taft for the Republican nomination was a spectacular victory of an attractive national and world celebrity, supported, too, by the most effective advertising techniques afforded by Madison Avenue. Eisenhower, the biggest of "names," won over a lesser "name." Eisenhower represented all of the new mass forces in politics; Taft represented all of the old and orthodox methods of the traditional American political game, even to the rounding up of delegates from the rotten-borough Republicans of the South in the gamy way sanctioned by age. On the Democratic side, Stevenson appeared to be the reluctant candidate seeking to avoid becoming the national favorite, but actually by the time of the Democratic convention he had been the subject of such a terrific national publicity buildup that the polls showed him running ahead of those who had engaged in the primary contests.

In 1956, as we have seen, Eisenhower was still the same glamorous celebrity he had been for many years, and in addition he was backed by the enormous power and prestige of the presidency, and aided by the smooth television techniques of Robert Montgomery and the shrewd showmanship of James C. Hagerty. Stevenson, in many a slugging primary, had pushed his way into first place among the Democrats. And, most significant of all, the trend in both parties toward nominating for President *the* national favorite appeared for the first time to be extending itself to the vice-presidential candidates as well.

IV

The true significance of what has been happening to our nominating conventions beginning with 1928 can be seen only when contrasted with the situation before 1928. Throughout most of American history, the nomination of a "dark horse" was not infrequent, and the nomination of a favorite son (a politician prominent in his home state, but not nationally) or of a national favorite who was not *the* national favorite was the common practice. Indeed, it has been only in the past thirty years that we have developed the amorphous but none the less fairly accurate methods of determining *the* national favorite in each party and the mass pressures which are now applied to produce his nomination.

In the past, we could speak accurately of *the* national favorite only when a popular sitting President was seeking renomination; or when there arose a magnetic personality with a passionately devoted personal following among the rank and file—a Thomas Jefferson, an Andrew Jackson, a Henry Clay, a Stephen A. Douglas, a William Jennings Bryan, or a Theodore Roosevelt; or, more rarely, in the case of a genuinely popular military figure like Ulysses S. Grant. Most military heroes nominated for the presidency have been synthetic political figures manufactured for the occasion by behind-the-scenes professional politicians—as in the nominations of William Henry Harrison, Zachary Taylor, Winfield Scott, John C. Frémont, and Winfield S. Hancock. At the time of their *first* nominations, Lincoln, Cleveland, Bryan, and Wilson were only national favorites among other national favorites; none of these was *the* national favorite. Theodore Roosevelt did not become *the* national favorite among the rank and file of Republicans until after he had succeeded to the presidency. The traditional pattern was clear: most conventions had before them a number of national favorites and favorite sons, and by a process of attrition a national favorite (usually not *the* national favorite, for *the* national favorite rarely existed) finally emerged victorious, or the contest between national favorites became so bitter that the convention was deadlocked and out of stalemate a favorite son, or more rarely, a "dark horse" emerged. (I put the term "dark horse" in quotation marks because actually a "dark horse" was usually a favorite son primed for entry at the strategic moment.)

The traditional way of nominating presidential candidates very frequently led to a miscarriage of the hopes and expectations of the rank and file of party members. Clay's loss of the nomination to Harrison saddened most Whigs, and Scott's nomination in 1852 left most Whigs cold. Among Democrats for the same period, Polk in 1844 was a dark horse; Cass in 1848 was a lesser national favorite; and Pierce in 1852 was another dark horse. In 1856, the Republican Frémont was a more or less synthetic figure, and the Democrat Buchanan represented the triumph of the professional politicians over Stephen A. Douglas, the favorite of the rank and file.

In the decades following the Civil War, the situation was much the same. The Democratic nominee of 1868, Horatio Seymour, was a favorite son. In 1872, the Democrats reluctantly endorsed the Liberal Republican nominee, Horace Greeley, whose nomination by the Liberal Republicans had occasioned amazement bordering on consternation. The Democratic nominee in 1880, General Hancock, was a somewhat ridiculous figure without popular appeal. On the Republican side, the nomination of favorite son Hayes in 1876 gravely disappointed the idolatrous followers of James G. Blaine, who was the national favorite with the rank and file. Garfield, the Republican nominee in 1880, was a dark horse; and Benjamin Harrison, the Republican nominee in 1888, was a favorite son.

During the first decades of the twentieth century, the nominating conventions began responding more to national and democratic forces outside the conventions, and these decades may be considered transition decades to the period which began in 1928. Nevertheless, many conventions of the first two decades of this century disappointed the democratic expectations of the country. The nomination of Alton B. Parker by the Democrats in 1904 was a "dud." The nomination of James M. Cox in 1920 represented the victory of a favorite son, or at best of a lesser national favorite. The nomination of favorite son John W. Davis in 1924 was the result of the prolonged deadlock in the Democratic convention and was distinctly anticlimactic. It came when a locked convention had taken 103 ballots and could not produce a more satisfactory candidate. On the Republican side, the nomination of President Taft over Theodore Roosevelt in 1912 clearly represented a victory of maneuver-by-politicians over the newer democratic methods; and in 1920 the nomination of Harding over three national favorites represented the triumph of local politicians and congressional cabal over the verdict of the primaries, the triumph of the old methods at their very worst.

V

Why have our nominating conventions been going through a process of revolutionary transformation during the past thirty years? Basically, the transformation is a response to the more integrated mass society in which we live. Increasingly we are dealing with problems that are remote and beyond personal experience, that appear in the guise of historical necessity or inevitability or crisis. The professional politician steeped in localism (and this includes the average congressional politician) now is less sure of himself, often feels his inadequacy, and turns to celebrities on the national and world stage for leadership. Hoover, Willkie, and Eisenhower became national celebrities outside politics before they became political leaders. Franklin Roosevelt's third and fourth nominations were made possible (indeed, "necessitated") by world crisis. (In the future, situations like these will increase, not decrease, and they likely will lead to an alteration of the constitutional amendment prohibiting more than two presidential terms.)

Increasingly, national leaders are being made in this way: the ambitious person seizes and exploits his "chance" on the national or international stage (as Kefauver did as Senate "crime-buster"), the mass media of publicity make him a "name," and, having become a "name," he is made a bigger "name" by the same mass media of publicity. We are approaching a condition where celebrities outside of politics—Hollywood, television, and radio stars, sports heroes, and fiction writers—or even the wives of celebrities (they bear the "name," do they not?) carry greater weight in political campaigns than do long-time congressional leaders or state governors.

The preconvention primaries and campaigning have done much to alter the nature of the nominating convention. True, the number of states having presidential primaries of one kind or another (to elect delegates to the national convention, or show a popular preference, or both) has not increased in the past thirty years, but these primaries are coming to have a significance in the nominating process they did not have in 1912 or 1920 or 1924. Today, when a politician enters the primaries and makes good showings, he becomes the subject of concentrated national publicity. His activities are carried every day by the national press services, and his name regularly appears in all the daily papers of the country. Human-interest stories about him are featured in the magazines of mass circulation. He appears on national radio and television programs. On a

Sunday afternoon he "meets the press" for millions to see and hear. National public-opinion polls carry his name and periodically rate his popularity. He becomes a "name," a celebrity. Even his wife or near relative becomes a "name"—Pat, Nancy, and "Buffie" become almost as well known as Nixon, Kefauver, and Stevenson. If put to the test, does anyone doubt that during the campaign of 1956 a Pat Nixon or a Nancy Kefauver could have drawn larger audiences than a Sam Rayburn or a Joe Martin?

The primaries allow the politician an opportunity to become a "name." They allow the unknown and the semiknown to become really known. They allow him the chance to become a celebrity along with the celebrities in other fields of activity and achievement. The time is fast approaching, if indeed it has not already arrived, when a politician cannot hope to be President or even vice president without first becoming a celebrity. The days when the political lightning in a nominating convention would strike an unknown or a comparative unknown are over. *If politicians do not find the ways to make themselves household names—real national celebrities—presidential nominations will go to celebrities in other fields.* The art of being an effective national political leader is compounded of many ingredients; but today the first ingredient is to know how to become a national celebrity, for without this the other ingredients cannot be put to use. For this reason, even if all the state presidential primaries were abolished, the skillful politician in search of the presidency, the politician not yet a national "name," would still have to go to the country, line up the mass pressure groups, beat the bushes, and corral delegates in a popular way. Only by these methods could he get the concentrated national publicity that makes him a celebrity. Without the primaries, his road to national fame would be still harder, but he would have to find it. That day is now passed when an astute politician can reach the presidency largely by the quiet search for delegates and the lining up of congressmen, local leaders, and bosses; largely by using the methods of closet conference, maneuver, manipulation, and bargaining.

Another reason for the transformation of the nominating convention is the increasing importance of mass pressure groups. Today, the candidate must not only deal with the leaders of the mass pressure groups, but he must also go to the members of these groups and convince them of his sincerity and ability to perform. Truman got the Democratic vice-presidential nomination in 1944 when he was found to be satisfactory to the big labor organizations. That year, Byrnes failed in his quest for the same nomination when national labor, Catholic, and Negro groups turned him down. Barkley's 1952 "boom" for the presidential nomination collapsed when it was punctured by the national labor leaders. Kefauver owed his 1956 vice-presidential nomination in no small part to the support of national farm and labor organizations.

It is curious that the transformation of the nominating conventions has not been more generally recognized. After all, the nomination of presidential candidates is as important as the election itself. The effects of an increasingly integrated and mass society have been recognized in almost all other areas of our life. We even recognize these in our presidential *elections*. It is pretty well understood that presidential elections are not longer won by exploiting local and state disaffections and patching together local interests and thereby eking out a national victory, as Garfield did in 1880, Cleveland in 1884, Harrison in 1888, and Cleveland again in 1892. (There is increasing group politics—and our parties are federations of many group interests—but it is more and more *national* group politics, and less and less local, state, and sectional politics.) There have been fourteen presidential elections during the twentieth century, and all of these, except those of 1916 and 1948, have been national landslides, expressions of national waves of opinion. However, we still persist in thinking of the presidential nominating machinery in terms of the parochial and provincial politics of the nineteenth century. Such thinking, while picturesque, is no longer realistic.

The activities that surround a nominating convention are deceptive. The proceedings and ritual on the floor of the convention itself are very much as they were in 1868 or 1880 or 1900. Outside, in the lobbies and hotel rooms, there is still the buttonholing of delegates, the anxious bargaining, the exciting secret conferences. But all of this is now subordinate in decision-making to the group and mass pressures, developed or enormously expanded and intensified during the past thirty years, which now beat upon the delegates and conferees from all sides.

Most Americans, even American political scientists, think in terms of form and structure. It is the heritage or our living under a written constitution. Because the forms and structure of the national nominating convention are little changed, we have been slow to recognize the revolution that has already been worked in it. The vast changes in the nominating practices of the localities and states are recognized because those changes were accompanied by changes in the machinery, in the abolition or drastic alteration of the old state-convention systems and the adoption of complete or mixed primary systems in the states. It was thought that similar changes in machinery would be applied to the national nominating conventions, that more states would adopt presidential primaries, or that a national presidential primary would be adopted. However, few structural and formal changes have taken place on the national level. There has been no increase in the number of presidential primary states in the past thirty years. The national primary has not been adopted, and it is not likely to be adopted.[1] A few states have clarified their presidential primary laws. One of these is Florida, which now requires party voters to select from slates of delegates pledged to particular candidates, just as presidential electors are now pledged to vote for the national nominees of their party. Hidden from national view, but significant, is the fact that in many of the states the delegates to the state conventions, which in turn select the delegates to the national conventions, are now chosen in local primaries.

The trend to nominating *the* leading national favorite, then, is the result of many factors: the crisis world in which we live; the rapidity with which events that affect the increasingly interrelated lives of all of us occur; the phenomenal growth of the mass media of communications; the increasing activities of mass pressure groups; the pervasiveness, accuracy, and notoriety of the public-opinion polls; the democratization of the methods of selecting delegates to the national conventions; and the ways all of these in turn intensify the importance of the presidential primaries. However, today, the democratic process is coming to depend less and less upon the formal and periodic nominating and electoral machinery and more and more upon the ever-multiplying means for informally and pervasively ascertaining public opinion from day to day. In short, the revolution has been wrought not so much by changes in machinery and structure as by the vast changes in social forces outside of machinery and structure, new and altered forces which operate on and permeate machinery and structure.

It is probable that by 1976 or 1980 all that a nominating convention will do will be to meet to ratify the nomination for President of *the* national favorite already determined by the agencies, formal and informal, of mass democracy; to ratify the nomination for vice president of the second leading national favorite; to endorse a platform already

1. If the trend toward democratizing the national convention should fail to continue, a new agitation for a national primary will emerge, particularly if the method of electing the President is democratized by abolishing the electoral college and substituting direct popular election of the President.

written by leaders responding to national and group pressures; and to stage a rally for the benefit of the national television audience. Delegates and "leaders" in national conventions, like presidential electors, will have become rubber stamps.

VI

These trends raise many specific problems in government and politics. I shall suggest merely a few of these.

For one thing, since the forces of national mass democracy are affecting the presidency much more and at a faster pace than they are affecting the Congress, we are now more and more faced with a widening gap between the President and Congress, between presidential politics and congressional politics, between presidential Republicans and congressional Republicans, between presidential Democrats and congressional Democrats. The implications of this for a frightening prolongation and intensification of the stalemate between the legislative and executive branches and for continued inability to enact needed legislation are obvious. What can be done about this is another story, and a complicated one. Primarily we must rely on the substantive realignment of parties now gradually taking place, on progressive Republicans in the prairie states and in the Northwest becoming Democrats, and on the conservatives in the South becoming Republicans. Premature attempts to hasten this trend by centralized national party machinery over congressional nominations and elections might actually retard the trend and aggravate the situation.

For another thing, are we not in danger of prematurely killing the chances of genuinely first-class national leaders and depriving ourselves of their services by overreacting nationally to parochial situations in individual and often provincial states? Why should international-minded Wendell Willkie have been broken on a single throw of loaded dice in a primary in an isolationist state like Wisconsin? Why should a Stevenson or a Kefauver be "made" or "lost" by the results of a primary in a state as atypical as Florida, where the political divisions normal to most states mean so little that its two United States Senators, representing the overwhelming middle-class consensus of its *rentier* classes, tourists, and comfortable civic-club and chamber-of-commerce communities, are returned to Washington with no opposition? (Fortunately for the survival of both Stevenson and Kefauver, in the 1956 Florida primary they ran neck-and-neck among the small number of voters who bothered to vote.)

Of most immediate concern, it seems to me, in the rescuing of our politicians from the intolerable wear and tear of their bodies and nervous systems which the requirements of mass democracy increasingly are imposing upon them. This is coming to be a question of their sheer physical survival, and as a nation we are running greater and greater risks in allowing our parties to be led during an entire election year— during the long months of preconvention campaigning and then during the months of the general election campaign itself—by harassed and exhausted candidates with little time to analyze for themselves the enormous problems with which they must cope.

How can our presidential candidates be rescued from the increasing brutalities and tortures of the system? Mass democracy will have to find ways to use pervasively and effectively the mass media of communication—television, radio, and motion pictures. These mass media are partly responsible for the fact that a candidate for President must now become a national celebrity in order to get a nomination. These mass media now make it possible for the candidate to reach the voters without going directly to them in person. Yet up to now the uses of these mass media have not been within adequate reach of the candidates, so they have enormously expanded their personal contacts and appearances, expanded them to the near-breaking point. This constitutes a cruel and preposterous paradox.

To meet the expanded costs of the preconvention contests and the campaign itself the federal government may have to step in to underwrite the expenses of radio and television campaigning and of a reasonable amount of newspaper and magazine advertising. Earlier in our history, governments did not even defray the expenses of preparing the ballots. These expenses and some other election costs were borne by the candidates and the parties. Now all of these things are regarded as the legitimate government costs of an election and are borne by the local and state governments. However, the costs of campaigning are as necessary as the costs of the election itself, and the public may have to come to accept public financing of reasonable campaign costs as legitimate, just as legitimate as the public financing of ballots, voting machines, polling places, and election clerks.

After this hurdle has been cleared, there will be all sorts of specific and technical difficulties. How are the bona fide candidates to be distinguished from the frivolous ones? Is a candidate who does not file in any of the state presidential primaries to be considered a candidate? Is the candidate who files in the state primaries to be considered a federal or a state candidate, and if he is to be considered a federal candidate for purposes of certain campaign costs, in how many states must he file? These are just a few of the questions that will arise.

During the first two decades of the twentieth century, the presidential election was subjected to the ways of modern mass democracy. We have not yet solved many of the problems involved in this transformation, particularly those involving campaign finances and adequate use of radio and television time. During the past three decades, the presidential nominating process likewise has been subjected to the ways of mass democracy. This constitutes a more decisive break with the past, for it involves a revolutionary transformation of the nominating conventions, a substitution of mass choices for the deliberations and decisions of political leaders, the subversion of the only effective institution that has stood between the presidency and mass pressures. This revolution is already pretty much of an accomplished fact, although that fact has hardly come to be recognized. It is high time that we recognize it; more, that we begin to tackle the problems which have been raised by this revolution.

๙ะ Kennedy in History: An Early Appraisal

The November 20, 1964 issue of *Life* magazine called the following article "a superbly balanced critique" of John F. Kennedy. On the other hand, Professor James Tracy Crown, in his book *The Kennedy Literature* (New York: New York University Press, 1968), characterized this article and another of mine, "The Cult of Personality Comes to the White House" (*Harper's Magazine*, December 1961), as weighted on the critical side. The author himself, rereading the following article after the passing of years, would not wish to change its over-all impression.

As a result of the Democratic presidential and congressional landslides of 1964 (which took place after the article below was written), the legislative logjam in Congress, referred to in the article, was finally broken, and a large and constructive program of liberal legislation, comparable to the Wilsonian and FDR programs, was enacted. However, in spite of President Lyndon Johnson's superior ability to deal with Congress, as compared with Kennedy's, this achievement did not so redound to the reputation of Johnson as to denigrate Kennedy, as this article anticipated. For, ironically, it was Barry Goldwater more than anyone else who was responsible. By taking so bold a reactionary stand during the presidential campaign of 1964, Goldwater so alarmed middle-of-the-road voters, including many Republicans, that they sent enough additional liberals to the Congress of 1965–1966 to overcome the conservative Republican–Southern Democratic coalition there and get liberal legislation moving at last.

In the light of President Johnson's subsequent escalation of the Vietnam war, the brief reference in this article to the Kennedy policy in Vietnam would now need elaboration.

After the passing of the years, what can be said now of the collective Kennedy legend? Will it survive? No. Will the romantic personal legend of John F. Kennedy survive? Yes, long after the sheen has worn off all the supporting cast.

■ ALTHOUGH John F. Kennedy has been dead less than a year, ▪t is not too early to assess his place in history. Judgments about historical

The Antioch Review, XXIV (Fall 1964).

figures never come finally to rest. Reputations fluctuate through the centuries, for they must constantly do battle with oblivion and compete with the shifting interests and values of subsequent generations. Even so, contemporary estimates are sometimes not markedly changed by later ones.

The career of a public man, the key events of his life, most of his record, the questions he confronted, the problems he tackled, the social forces at work in his time are largely open to scrutiny when he leaves public life. The archives and subsequently revealed letters, memoirs, and diaries will yield important new material, but usually these merely fill out and embellish the already known public record.

What more frequently alters an early historical appraisal than later revealed material is the course of events after the hero has left the stage. For instance, should President Johnson, during the next few years, succeed in breaking the legislative logjam and driving through Congress a major legislative program, then President Kennedy's failure to do this will be judged harshly. But should the congressional stalemate continue, then the Kennedy difficulties likely will be chalked up not so much against him personally but against the intricacies of the American system of separation of powers and pluralized, divided parties. Even so, it will not be easy to defend the meager Kennedy legislative performance.

If popular passion is running high at the time a historical personage makes his exit, then it is difficult for a contemporary to be relatively objective in his evaluation. (At no time is there absolute objectivity in history.) For instance, Woodrow Wilson left office amidst such violent misunderstanding that few contemporaries could do him justice, and only now, four decades later, is he being appraised with relative fairness. With Kennedy, one collides with adoration. The uncritical bias in favor of Kennedy derives from his winsome personality, his style and élan, the emotional euphoria arising from his tragic death, and the sympathetic spiritual kinship with him felt by historians, political scientists, intellectuals, and writers. There is, to be sure, considerable hate-Kennedy literature, most of it written before his death and reflecting ultra-right wing biases, but few will be fooled into taking diatribes for serious history.

II

Although as President he clearly belongs to the liberal tradition of the Democratic party's twentieth-century Presidents, Kennedy's pre presidential career can scarcely be said to have been liberal. During his congressional service he refused to hew to an ideological or even a party line. Kennedy often asked: "Just what is a liberal?" He confessed that his glands did not operate like Hubert Humphrey's.

Reviewing Kennedy's votes in House and Senate from 1947 to 1958, the AFL-CIO's Committee on Political Education gave him a pro-labor score of twenty-five out of twenty-six, which in considerable part may be explained by the large number of low-income wage-earners in his district. Kennedy usually followed the views and interests of his constituents, and this gave his congressional career its greatest measure of consistency. With respect to welfare programs which did not much concern his constituents—farm price supports, flood control, TVA, the Rural Electrification Administration—Kennedy sometimes joined the Byrd economy bloc. On occasion, Kennedy could fly in the face of New England sentiment, as when he supported reciprocal tariff laws and the St. Lawrence Seaway, and again when he publicly stated that he thought President Eisenhower, in an attempt to salvage the Paris summit conference, should apologize to Khrushchev for the U-2 incident. At such times, Kennedy displayed a disarming candor.

In foreign policy, Kennedy's early pronouncements had a right-wing flavor, especially with respect to China, but he soon developed a greater interest in foreign affairs than in domestic ones, and he moved away from right-wing attitudes to favor foreign economic aid, to rebuff the Committee of One Million devoted to keeping Red China out of the United Nations, to speak up for abandoning Quemoy and Matsu, to fight for economic help to Communist Poland, to advocate Algerian independence, and to encourage international arms-control negotiations.

In general, through the years, in both domestic and foreign policy, Kennedy moved from positions right of center or at center to positions left of center.

Since Negro rights played so important and dramatic a part in his presidency, historians will always be interested in Kennedy's earlier stand on civil rights. Until the presidential campaign of 1960, Kennedy was not an aggressive fighter for the Negro. During his quest for the vice-presidential nomination in 1956, he wooed the Southern delegations and stressed his moderation. After 1956, he sought to keep alive the South's benevolent feeling for him, and his speeches in that region, sprinkled with unflattering references to Reconstruction "scalawags" like Mississippi Governor Alcorn and praise for L. Q. C. Lamar and other Bourbon "redeemers," had a faint ring of Claude Bowers's *Tragic Era*.[1] In 1957, during the fight in the Senate over the civil rights bill, Kennedy lined up for the O'Mahoney amendment to give jury trials to those held in criminal contempt of court. Civil rights militants regarded this as emasculating, since those accused of impeding Negro voting would find more leniency in Southern juries than in federal judges.

What will disturb historians most about Kennedy's pre-presidential career was his evasion of the McCarthy challenge. In his early congressional days, Kennedy himself often gave expression to the mood of frustration, especially with reference to the loss of China, out of which McCarthyism came. In a talk at Harvard in 1950, Kennedy said that "McCarthy may have something," and he declared that not enough had been done about Communists in government.[2]

By the time McCarthyism had emerged full blown, Kennedy was caught in a maze of entanglements. His Irish-Catholic constituents were ecstatically for McCarthy, and Kennedy's own family was enmeshed. Westbrook Pegler, McCarthy's foremost journalistic supporter, was close to father Joseph P. Kennedy; the elder Kennedy entertained McCarthy at Hyannis Port and contributed to his political fund; brother Robert Kennedy became a member of McCarthy's investigating staff. John Kennedy himself not only kept mum on the McCarthy issue but actually benefited from it. During Kennedy's senatorial campaign against Lodge, the latter was portrayed as soft on Communism, leading McCarthyite politicians and newspapers in Massachusetts backed Kennedy, and McCarthy was persuaded to stay out of Massachusetts and do nothing for Lodge there.

When McCarthyism finally climaxed in the debate and vote in the Senate over censuring the Great Inquisitor, Kennedy was absent because of an illness which nearly cost him his life. At no time did Kennedy ever announce how he would have voted on censure. He did not exercise his right to pair his vote with another absent colleague who held an opposite view. It may be that Kennedy was too ill for even that. Kennedy's subsequent private explanations—that to have opposed McCarthy would have been

---

1. Kennedy made a number of speeches in the South in 1957. Illustrative of their tone is the commencement address at the University of Georgia and the address at the Democratic state dinner at Jackson, Mississippi.

2. See John P. Mallan, "Massachusetts: Liberal and Corrupt," *The New Republic*, October 13, 1952.

to commit political hara-kiri and that he was caught in a web of family entanglements—were lame and unworthy.[3]

Nevertheless, there can be no doubt that McCarthyism was a searing experience for John Kennedy. During his convalescence, as a sort of personal catharsis, he wrote his book *Profiles in Courage*, which dealt with famous senators of the past who had to choose between personal conscience and violent mass execration and political extinction. Kennedy seems to be saying: "This is the kind of cruel dilemma I would have faced had I been present in the Senate when the vote was taken on the McCarthy censure." *Profiles in Courage* is Kennedy's examination of his own Procrustean anguish. Such soul-searching may have helped him psychologically, but it added not a whit to the fight against McCarthy's monstrous threat to the free and open society.

Some years later, when even then Eleanor Roosevelt sought to prod Kennedy into making an ex post facto condemnation of McCarthyism, he wisely refused. He observed, with that candor which was his most engaging hallmark, that since he had not been in opposition during the controversy, to take a decided stand after Senator McCarthy was politically dead would make him (Kennedy) a political prostitute and poltroon.[4]

In the future, historians sympathetic to President Kennedy are likely to underestimate seriously the right-wing influences that played upon him and his brother Robert during their formative years and early life in politics. The influence of Joseph P. Kennedy was strong, and in the pre–Pearl Harbor years the Kennedy household was the center of "America First" leaders, journalists, and ideas and in the early 1950s of McCarthy leaders, journalists, and ideas. It is not remarkable that John and Robert reflected some right-wing attitudes in their early political careers; what is remarkable is that first John and then Robert, who is less flexible, emancipated themselves from these attitudes.

III

In all likelihood, John Kennedy would never have become President had his brother Joe lived; or had he (John) been nominated for vice

3. See an interview granted by Senator Kennedy to Irwin Ross, the New York *Post*, July 30, 1956.
4. For the Kennedy-Eleanor Roosevelt byplays over McCarthy, including the incident here mentioned, see Ralph G. Martin and Edward Plaut, *Front Runner, Dark Horse* (Garden City, N. Y.: Doubleday, 1960), pp. 74–75; J. M. Burns, *John Kennedy: A Political Profile* (New York: Harcourt, Brace & World, 1960), p. 153; Eleanor Roosevelt, *On My Own* (New York: Harper & Row, 1958), pp. 163–164; Alfred Steinberg, *Mrs. R.: The Life of Eleanor Roosevelt* (New York: G. P. Putnam, 1958), p. 343.

president in 1956; or had the Paris summit of May 1960 not blown up.

All of those who have followed the Kennedy story know well that it was the eldest son of the family, Joe Jr., and not John, who was slated for politics. If Joe Jr. had lived, John would not have gone into politics at all. This is not to say that Joe Jr. would have "made the grade" in high politics, as believers in the Kennedy magic now assume. Joe Jr. was an extrovert; he was obviously the political "type." John's mind was more penetrating and dispassionate, and he did not fit the stereotype of the politician, particularly the Irish politician. What endeared John to the status-seeking minorities was that he appeared more the scion of an old aristocratic Yankee family than the authentic scions themselves. Had Joe Jr. lived, the Kennedy family in all probability would never have had a President at all.[5]

Had John Kennedy been nominated for vice president in 1956, he would have gone down to disastrous defeat along with Stevenson. That year Eisenhower carried even more of the South than he had in 1952. Both the elder Kennedy and John are reported to have felt that this poorer showing by the Democratic national ticket in 1956 would have been attributed to the continued vitality of anti-Catholic sentiment, thus rendering Kennedy unavailable for the 1960 nomination. It seems that this is a sound judgment.

Despite the fact that more Americans were registered as Democrats than as Republicans, the chances for Republican success in 1960 looked rosy indeed until after the dramatic U-2 incident and the collapse of the Paris summit conference. Had the summit accomplished anything at all, the result would have been hailed as the crowning achievement of Eisenhower, the man of peace, and Nixon would have been effectively portrayed as Eisenhower's experienced heir, one who could be both resolute

5. In part, this evaluation of Joe Jr. and John is derived from personal observation. I recall vividly an evening, April 4, 1941, when I was a guest at the Kennedy home in Palm Beach. Following dinner, the entire family, including the younger children, assembled in the drawing room for a discussion of public affairs. Ambassador Kennedy was particularly interested in canvassing our views about the content of commencement addresses he was to deliver later that spring at Oglethorpe University and at Notre Dame University, but the discussion ranged widely over world politics and foreign policy. Mr. Kennedy, John, and I were the chief participants, although Mrs. Kennedy and Joe Jr. often broke in with comments. It was clear to me that John had a far better historical and political mind than his father or his elder brother; indeed, that John's capacity for seeing current events in historical perspective and for projecting historical trends into the future was unusual. After the family conclave broke up, John and I continued an animated conversation into the early morning hours.

and conciliatory. The explosion of the summit paved the way for Republican defeat and gave plausibility to the kind of "view-with-alarm" campaign waged by Kennedy. In any general climate favoring Republican victory, even Kennedy's ability to win the Catholic vote would not have elected him.

Will historians be hard on Kennedy for his extravagant charges during the 1960 campaign that the Eisenhower administration had been remiss about America's missiles and space programs? (The fact is that the "gap" in intercontinental missiles was decidedly in favor of the United States.) No, for Presidents are rated in history by the records they make in office, not by how they wage their campaigns. In his first campaign for the presidency, Lincoln assured voters that the South would not secede; Wilson promised his New Freedom but in office wound up closer to Theodore Roosevelt's New Nationalism; and Franklin Roosevelt pledged economy and a balanced budget but became the founder of the New Deal. Kennedy's zeal will likely be put down as no more than excessive "campaign oratory." Oddly enough, it may be Eisenhower who will be blamed for failing to "nail" Kennedy and to tell the nation bluntly the true state of its defenses, for missing the opportunity to add credibility to America's nuclear deterrent. Liberal historians, impressed by Kennedy's domestic promises and, except for Cuba, his more liberal program than Nixon's in foreign policy—Quemoy and Matsu, emphasis on foreign economic aid, stress on international arms control—will not be prone to hold Kennedy accountable for the alarmist flavor of his campaign.

Of 1960's "issues," only the religious one will loom large in history. It was the religious implication which gave Kennedy the victory. Democratic gains in the pivotal Catholic cities in the states with the largest electoral votes were greater than Democratic losses in the Protestant rural areas. It was the Democrats, not the Republicans, who actively exploited the religious aspect. In the Protestant areas, citizens were made to feel they were bigots if they voted against Kennedy. But Catholics were appealed to on the grounds of their Catholicism. Any evidence of Protestant intolerance was widely publicized to stimulate Catholics and Jews to go to the polls to rebuke the bigots. But history will treat these Democratic tactics—a kind of bigotry in reverse—with kindness. The means may have been objectionable, but the good achieved was enormous. For the first time in American history a non-Protestant was elected President. The old barriers were downed. Now that an Irish Catholic had been elected President, the way was opened for the election of an Italian Catholic, a

Polish Catholic, a Jew, eventually a Negro. A basic American ideal had at last been implemented at the very pinnacle of American society.[6]

The second permanent contribution of the 1960 election lies in its underscoring the large degree to which presidential preconvention campaigns and election campaigns have been geared to democratic mass behavior. Kennedy and his team wisely recognized that a mere pursuit of the politicians and delegates was not enough, that beginning with 1928, conventions had nominated the outstanding national favorite as indicated by the primaries and the polls. In Kennedy's case, winning the primaries and the polls was especially necessary to convince the doubters that a young man and a Catholic could be elected President. Hence Kennedy organization, money, and high-level experts were directed not merely to bagging delegates but even more to winning mass support, primaries, and high ratings in the polls. In this they succeeded marvelously. In effect, the Democratic convention merely ratified the choice already made by the primaries, the polls, and the mass media. The revolution in the presidential nominating process had been in the making for over three decades, but it took the Kennedy campaign to make the public and even the pundits aware of it.

During the election campaign itself, Kennedy kept alive his personal organization; brother Bob, his personal manager, was more important than the chairman of the Democratic National Committee; the nominee himself made all the meaningful decisions and virtually monopolized the limelight; other party leaders were dwarfed as never before. The TV

6. In emphasizing the religious issue as affecting the outcome of the campaign, I incline to the conclusions reached by Elmo Roper in "Polling Post-Mortem," *Saturday Review*, November 26, 1960; the Gallup figures in "Catholics' Vote analyzed, a 62% Switch to Kennedy," New York *Herald-Tribune*, December 6, 1960; Richard M. Scammon, "Foreign Policy, Prestige Not a Big Election Factor," Washington *Post*, December 15, 1960; Louis Bean comment in Anti-Defamation League of B'nai B'rith press release, January 14, 1961. These coincide with my own findings as the result of interviews with voters in certain sensitive election districts during the campaign of 1960. However, behavioral statisticians differ about this, and doubtless a number of more definitive studies will be forthcoming. Whatever the reasons for the election of Kennedy, the fact that for the first time a Catholic was elected President is what has made the large and permanent impact on history. Even so, every Kennedy biographer must of necessity be concerned with the old question of means and ends, the moral implications of using a religious prejudice in reverse to break the old Protestant monopoly on the presidency. A striking example of appealing to religious emotion in the guise of opposing it is found in Robert Kennedy's Cincinnati address of September 13, 1960, which was climaxed by this: "Did they ask my brother Joe whether he was a Catholic before he was shot down?"

debates further spotlighted the nominees. Again, as in the preconvention campaign, the Kennedy team did not create the trend to a personalized campaign, to glamourous celebrity politics. The trend and the techniques to make it work had been on the way for decades. Basically these emerged from the increasingly mass nature of American society. But Kennedy exploited the trend and the techniques in conspicuously successful fashion; he widened, intensified, and accelerated them; he made the nation aware of them; he did much to institutionalize them.

Thus the Kennedy campaigns will always be remembered for the dramatic way they contributed to the personalized and plebiscitic presidency.

IV

No administration in history staffed the executive departments and the White House offices with as many competent, dedicated, and brilliant men as did Kennedy's. Kennedy paid little attention to party qualifications at top level; the emphasis was on ability, drive, imagination, creativity. Politicians made way for specialists and technicians; but Kennedy was on the lookout for specialists *plus*, for men who had not only technical competence but intellectual verve. Kennedy himself was a generalist with a critical intelligence, and many of the most prized of his staff were men of like caliber—Sorensen, Goodwin, Bundy. After the brilliance of the Kennedy team, and with the ever-growing complexity of government problems, no administration is ever likely to want to go back to the pedestrian personnel of earlier administrations, although few Presidents are apt to have the Kennedy sensitivity and magnetism capable of gathering together so scintillating an administration as his. FDR will be known as the founder of the presidential brain trust, but Kennedy will be known as the President who widened and institutionalized it.

In contrast to his performance in the executive departments, Kennedy's relations with Congress can scarcely be said to have been successful. The dream of enacting a legislative program comparable to that of Wilson and FDR soon vanished. The one outstanding legislative achievement of Kennedy was the Trade Expansion Act of 1962. All of Kennedy's other major goals—farm legislation, tax reform, a civil rights law, medicare, federal aid to schools—bogged down in Congress.

Since 1938, major welfare legislation had repeatedly been smothered in Congress at the hands of a Republican-conservative Southern Democratic coalition. During the bobtail, postconventions session of 1960, both Kennedy and Johnson, the Democratic party's new standard-bearers, had met humiliating legislative failure when they found themselves unable to budge the Democratic Congress. Kennedy had explained that once the powers of the presidency were in his hands, things would be different. But when he achieved the presidency and still failed with a Democratic Congress, Kennedy apologists contended that after Kennedy's re-election in 1964 he would be in a position to press more boldly for legislation. This argument stood experience on its head, for most Presidents have secured much more legislation during their honeymoon first years than during their lame-duck second terms.

Kennedy will not escape all blame for his legislative failures, for despite his awareness of the stalemate since 1938, he had promised a "strong" legislative leadership like that of Wilson and the second Roosevelt. Moreover, in various ways Kennedy contributed to the personalized and plebiscitic presidency: by the manner in which he waged his 1960 campaign, by his assigning to key posts not party leaders but men personally chosen for their expertise and creative intelligence, and by his monopolizing of the limelight. Kennedy and his family naturally made exciting publicity, but the President seemed to go out of his way to get even more—holding televised press conferences, for example, and permitting TV cameras to capture the intimacies of decision-making in the executive offices and of private life in the White House. All of this further exalted the presidency, further dwarfed politicians, party, and Congress, and added to Congress' growing inferiority complex.

Now, Kennedy was not unaware of the susceptibilities of Congress. He carefully cultivated individual congressmen and senators, frequently called them on the phone, had them up for chats, extended them an unusual number of social courtesies and parties in the White House. His legislative liaison team, headed by Kenneth O'Donnell and Lawrence O'Brien, was diplomatic and astute, pumped and twisted congressional arms, applied both the carrot and the stick. But Kennedy left too many of the congressional chores to his liaison team. He simply did not give this aspect of the presidency enough attention. Foreign affairs interested him intellectually much more than domestic measures. Despite his years in Congress and his love of politics, Kennedy did not really like or feel at home with small-bore politicians and congressional "types," and he was not skillful in his personal bargaining with them.

Moreover, Kennedy made no attempt to initiate and institutionalize new devices for easing presidential-congressional relations, nor did he even explore this problem intellectually. The breath of life to politicians is publicity, but no effort was made to share the presidential glory, of which there was a superabundance. What could be done to enhance the publicity and prestige of congressional leaders and committee chairmen who consented to carry the administration ball in Congress? How give them credit for "creating" and "initiating" the administration's legislative measures? How let them become spokesmen before the nation of the administration's legislative goals? True, other Presidents had made no such probings, but in recent years, with a legislative logjam piling up, the presidential-congressional deadlock had reached crisis-like proportions. The President did not give this question, inherently baffling at best, his full creative effort.

Kennedy's presidency will be known as the time of the Negro revolution, when Negro aspirations widened to include desegregation in the private sector and were spectacularly supported by sit-ins and street demonstrations. As President, Kennedy not only gave full executive backing to the enforcement of court decisions but personally identified himself with the goals of the Negro revolution and gave them the full moral support of the presidency.

By 1960, Kennedy had become an aggressive fighter for Negro rights. With the South lined up behind the Lyndon Johnson candidacy, Kennedy's nomination depended on the support of the Northern liberals and the metropolitan areas outside the South. During the election campaign, Kennedy's strategy was geared to winning the Negro vote in the big cities of the states with large electoral votes. Kennedy's new militancy carried over to his presidency.

But by 1963, it appeared that what had been a political advantage might turn into something of a political liability. "The Kennedys" were denounced in the South, and the President faced the loss of much of that section in 1964. More serious, there were indications that the civil rights issue would cost Kennedy many votes in the North, where considerable opposition to the Negro drive had developed. However, by this time Kennedy had chosen his course, and while there might be temporary shifts in tactics, there could be no turning back. Robert Kennedy has stated that at this point the administration really did not have any choice and that, besides, the administration's course was the correct one. He reports the President as saying: "If we're going to lose, let's lose on principle."

There seems little question that Kennedy would have been re-elected in 1964, but the civil rights issue would have been his biggest worry. In sizing up Kennedy as a politician, it is significant that he appears not to have anticipated the extent to which his position on civil rights might become politically hazardous. Otherwise it is difficult to explain the appointment of his brother as attorney general, upon whom the brunt of enforcing the civil rights court decisions would necessarily fall. Astute rulers take care to divert the political lightning of an offended public from themselves to subordinates. But in appointing his brother attorney general, the President left himself no "out." Those hostile to the Negro revolution could not say: "President Kennedy is all right; it is that attorney general of his." Instead, they blamed "the Kennedys." With another attorney general, President Kennedy might well have escaped some of the venom of the opposition. And incidentally, Robert Kennedy, in some other important job, would have been made better available for high politics in the future.

V

In foreign policy, the first two years of Kennedy were ambiguous. In the third year, there was a clearer sense of direction, one which promised to harmonize American policy with emerging new realities in the world.

At the time of the Kennedy accession, the postwar world was disintegrating. Bipolarization was giving way to depolarization. The Sino-Soviet rift was widening. With the single exception of little Vietminh, all the old European colonies that had recently gained their independence had escaped Communism, although there were Communist guerrilla activities in some of them. The trend was to a new pluralism, a new diversity. The nuclear revolution in war and the American-Soviet nuclear deterrents had rendered an ultimate military showdown unthinkable. The United States was ahead in the nuclear arms race.

In Europe, despite Khrushchev's bluster about West Berlin, the existing arrangements in East and Central Europe were ripening into a more overt *modus vivendi,* by way of tacit understanding rather than formal political agreements. Trade and intercourse between East and West Europe were increasing, the satellites were operating more independently of Moscow, and an all-European economic and cultural co-operation seemed slowly to be replacing the postwar's clear-cut division between the "two Europes." West Europeans were becoming less interested in NATO because they were more and more convinced that there would be no Soviet military aggression in Europe, due to the nuclear deterrent and other reasons. The drive to West European politicial integration was slackening, owing to the decline of external pressures and to DeGaulle's opposition to the supranational approach. Forces within the Six, composing the Common Market, were honestly divided over whether they wanted an inward-looking European community or an outward-looking Atlantic one.

In short, Kennedy was confronted with a new fluidity, a necessity and an opportunity for a reappraisal of American foreign policy. How much of the old foreign policy was still applicable? What aspects required a new orientation? To what degree was it safe, realistic, and advantageous to strike out in new directions? In some ways this ambiguous situation was more agonizing to decision-makers than the obvious crisis situation with which Truman and Acheson had had to deal in the late 1940s and early 1950s. It is no wonder that some aspects of the Kennedy record in foreign affairs seem somewhat confused, even contradictory.

The chief stumbling block to an American-Soviet *détente* continued to be Berlin, the two Germanies, and the territorial arrangements in East and Central Europe. Kennedy rejected explorations of a definitive settlement, and if in the future a genuine American-Soviet *rapprochement* develops, this rejection is likely to be held against him. However, he did move informally in the direction of a more openly tacit recognition of the existing arrangements in East and Central Europe. He deferred less to Adenauer's view than previous administrations had done. In his interview in *Izvestia*, remarkable for its clarity and candor, he agreed that it would not be advisable to let West Germany have its own nuclear weapons. After the Communists built the Berlin Wall, Kennedy resisted all pressures to use force to tear it down.

Nevertheless, during his first two years in office, Kennedy seems needlessly to have fanned the tensions of the dying Cold War. (It may be that "needlessly" is too strong a word; perhaps Kennedy thought he needed to arouse the country to obtain a more balanced military program, more foreign economic aid, the Alliance for Progress; perhaps he thought, too, that a truculent tone was necessary to convince Khrushchev that America would stand firm under duress for its rights in Berlin.) His inaugural address was alarmist, already historically off key, more suited to the Stalinist era than to 1961. His first State of the Union Message was even more alarmist. The nation was told that the world tide was unfavorable, that each day we were drawing near the maximum danger. His backing of the Cuban invasion in April 1961 further fanned the Cold War. His statement to newspaper publishers and editors gathered at the White House in May—that the United States was in the most critical period of its history—increased the popular anxieties. He overreacted to Khrushchev's

Vienna ultimatum in June, for in recent years Khrushchev's repeated deadlines and backdowns over West Berlin had become a kind of pattern. But for Kennedy, Vienna seems to have been a traumatic experience. On his return home he appealed to Americans to build do-it-yourself bomb shelters, and this produced a war psychology in the country and all manner of frenetic behavior, caused right-wingism to soar (1961 was the year the membership and financial "take" of the right-wing organizations reached their peak[7]), and weakened confidence abroad in Kennedy's judgment.

There are no defenders of the Cuban fiasco of April 1961. Even had the expedition of the Cuban exiles been given American naval and air support and forced a landing, there is scant evidence that the Cubans, at that time devoted to Castro, would have revolted en masse and welcomed the invaders as deliverers. More likely a nasty civil war would have followed, with the Americans, giving increasing support to the invaders, cast in the role of subjugators. The C.I.A. had already rejected the social-revolutionary leadership of the anti-Castro Manuel Rey for a nonleftist leadership, and this would have made the task of overthrowing Castro even more difficult. The world would have looked on with dismay, and outside the United States the whole affair would have come to be regarded as "another Hungary." It is ironical that Kennedy, the generalist with a critical intelligence, the politician with a feel for popular moods, should on this occasion have been taken in by the bureaucrats and the "experts." Prodded by his own anti-Castro stand during the election campaign, Kennedy must have wanted desperately to believe in the reliability of those dossiers of the intelligence agents.

With respect to Western Europe, the Kennedy administration underestimated those forces within the Common Market that wanted a European community rather than an Atlantic community, at first regarded De Gaulle as a kind of maverick without group support for his position, and framed the Trade Expansion Act of 1962 in such a way that the most decisive tariff cuts between the United States and the Common Market would depend upon Britain's inclusion in the Market. Nevertheless, the Act as written still allowed for much liberalization of trade, even with Britain outside the Market, and the responsibility for failure to take advantage of this opportunity must be borne by parochial minded groups and interests inside the Market.

7. Donald Janson and Bernard Eismann, *The Far Right* (New York: McGraw-Hill, 1963), pp. 56 and 127.

The Kennedy administration's contributions to national defense were notable. It emphasized a balanced and diversified establishment—both strategic and tactical nuclear weapons, conventional arms, and guerrilla forces—so the nation would never have to make the choice between the ultimate weapons and no other adequate defense. It was realistic in its shift from bombers to missiles as the chief nuclear carriers of the future, and in its dismantling of the intermediate-missile bases in Britain, Italy, and Turkey as the Polaris submarines and intercontinental missiles became increasingly operational. Its attempt to find a formula for a NATO multilateral nuclear force was a way of countering De Gaulle's blandishments to the West Germans and of balancing the possibility of a *détente* with Russia with reassurances to Bonn. Its experiments with massive air-lifts of ground troops was in part a response to the desires of many of America's NATO allies for less rigidity, less insistence on fixed ground quotas, and more flexibility. However, NATO was plainly in transition, and while the Polaris submarines and intercontinental missiles were making the United States less dependent on European bases, ways were not yet actually implemented to share America's nuclear weapons with European allies on a genuine multilateral basis and satisfy their desires for less centralized direction from the United States.

There was an honest facing up to the terrible responsibilities inherent in the nuclear deterrent. That deterrent was put under tighter control to guard against accident and mistake, and the "hot line" between Washington and Moscow was set up. A much more determined effort was made to get arms-control agreements and a treaty banning nuclear-weapons testing than had ever been made by Kennedy's predecessors. Negotiations with the Soviet Union had been going on for years, but the Americans now so yielded in their former demands for strict international inspection as to put the Russians on the defensive, making world opinion for the first time believe that it was the Russians and not the Americans who were the obstructionists. Kennedy's administration believed that the United States and Russia had an enormous common interest in preventing the spread of nuclear weapons to other countries, that the Sino-Soviet rift gave Khrushchev a new freedom and a new urge to make agreements, and that the increasing accuracy of national detection systems made the possibility of cheating on a test-ban treaty, even one without international inspection, "vanishingly small."

Kennedy's regime also showed its international-mindedness in its firm support of the United Nations. It defended the Secretariat, the executive, from Soviet attacks, and in practice the activities of the Secretariat were widened. The organization was saved from bankruptcy by American financial aid. The operation of the United Nations military force in the Congo, backed by the United States, showed that the American government had no sympathy for "neo-colonialism" as practiced by the Katanga secession, and it added another successful precedent for international enforcement of international decisions.

With respect to the underdeveloped nations, the Kennedy policies paralleled the trend of history. Anticolonialism and self-determination were more valiantly espoused than in the preceding administrations. The Dulles doctrine that neutralism is "immoral" was abandoned, and neutralism was cordially accepted for nations which wanted it. Neutralism was positively encouraged in Laos and in the Congo. Help to South Vietnam was so hedged as to prevent the guerrilla war there from escalating into another Indo-China war, another Korea. Foreign economic aid was increased. The Food-for-Peace program was expanded. The Peace Corps was launched. The Alliance for Progress, an ambitious economic-aid program in Latin America coupled with domestic reforms, an experiment in "controlled revolution," was undertaken.

However, Kennedy, like his predecessors, did little to make the average American understand foreign economic aid—that it is not only an attempt to raise living standards, prevent Communism, and contribute to the world's economic well-being and stability but is also a substitute for those obsolete ways in which the old colonialism supplied capital to the underdeveloped areas. Until an American President takes to television and in a series of fireside chats explains to Americans in simple terms the real meaning of the foreign-aid program, that program will be in jeopardy.

The Cuban crisis of October 1962, provoked by the discovery of secret Soviet intermediate missiles in Cuba, was the high point, the turning point, in the Kennedy administration. Could this crisis have been avoided? This will be debated by future historians. True, Khrushchev could not have declined giving Castro economic aid, technical assistance, and some military help, even had he desired to do so, for to have refused this would have been tantamount to surrendering Communist leadership to the Chinese. But why did he go to the length of planting intermediate-missile bases in Cuba? As an appeasement to the Stalinist and Chinese opposition? As a countermeasure to American missile bases in Turkey (which were soon to be dismantled)? As a means of blackmailing Americans into making a compromise on Berlin? To extract a promise from the

Americans not to invade Cuba? Whatever the causes, some future historians will have nagging questions: Might this terrible gamble in nuclear brinkmanship have been prevented had Kennedy previously shown more disposition to come to a *détente* with the Soviet Union by a somewhat clearer recognition of the two Germanies and other de facto boundaries and arrangements in East and Central Europe; and if so, did this Kennedy reluctance, coming in part out of regard for West German opinion, represent a realistic appraisal of the world situation?

Anyway, when the crisis came, even neutralist opinion seemed to feel that Khrushchev's attempt to compensate for his own intercontinental-missiles lag and the open and avowed American intermediate missiles in Turkey did not justify the sneaky Soviet operation in Cuba. America's quiet, deliberate planning of countermeasures, both military and diplomatic, was masterly. America's prudent use of force, enough but not more than enough to achieve its objective, won world-wide acclaim. Khrushchev and Castro lost face. The Chinese denounced the Soviet backdown, and Chinese-Russian relations worsened. Most important, the peak of the crisis, a spectacular nuclear brinkmanship, cleared the atmosphere like a bolt of lightning. The lunacy of an ultimate nuclear showdown was traumatically revealed. Khrushchev's personal correspondence to Kennedy, reputedly revealing a highly emotional state and a genuine horror of nuclear war, the President had the grace, sportsmanship, and wisdom to keep secret.

Thereafter Khrushchev spoke even more insistently about the need to avoid nuclear war and pursue a policy of peaceful but competitive coexistence. From then on Kennedy gave more public recognition to emerging new international realities, the world's escape from monolithic threats, the trend to pluralism and diversity. In his address at American University in June 1963, Kennedy spoke as if the Cold War scarcely existed and emphasized the common stake both the United States and the Soviet Union had in world peace and stability. This address, one of the noblest and most realistic state papers of our time, will be remembered long after Kennedy's innaugural address is forgotten.

The new spirit in world affairs expressed itself concretely in the consummation of the limited nuclear test-ban treaty in the summer of 1963, the first real break in the American-Soviet deadlock. After this, Kennedy proposed a joint American-Soviet effort to explore the moon, and he agreed to permit the Soviet Union to purchase American wheat.

By 1963, then, Kennedy had come to much awareness that the postwar world was ending and to a determination to attempt more shifts in American foreign policy in harmony with the emerging fluidity. By this time, too, he had developed close personal relations with a large number of premiers and heads of state the world over. It was felt that after his re-election in 1964 he would be in an unusually strong position to give American foreign policy a new direction, that the test-ban treaty was but a foretaste of more significant measures yet to come, measures which might lead to an American-Soviet *détente*, eventually even to a *rapprochement*. Thus the President's life ended in a tragic sense of incompleteness and unfulfillment.

Every twentieth-century American President with a flair for world politics and in power in time of momentous international decision has been felled by sickness or death before his term was over, before his work was completed. First Wilson. Then Roosevelt. Then Kennedy. For sheer bad luck, this is a record unique among nations.

VI

Because of the vividness of his personality and the shortness of his tenure, Kennedy will be known more for the intangibles—a taste-maker, a symbolic embodiment of the values of his time, a romantic folk hero—than for his achievements in statesmanship.

Government requires pageantry, and rulers are expected to put on a show. The Kennedys put on a superb one. Never before, not even under Dolly Madison, was the White House the scene of such a dazzling social life, one which combined beauty and intelligence, radiance and creativity. There were, to be sure, crabbed Mrs. Grundys who derided "peacock opulence" and looked back longingly to the decorous days of Lucy Webb Hayes. But most Americans were fascinated, pleased as punch that even Elizabeth and Philip appeared a bit dowdy in contrast to those two young American thoroughbreds in the White House. They figuratively crowned Jacqueline Queen of Hearts. This aspect of the Kennedy reign has been inimitably described by Katherine Anne Porter, and no historian will ever record it with more grace, insight, and tenderness.[8]

Kennedy's contributions to the cultural life of the nation also belong to the intangible, and they are difficult to measure. Now of course President Kennedy did not engage in as wide-ranging an intellectual life as President

8. Katherine Anne Porter, "Her Legend Will Live," *Ladies' Home Journal*, March 1964.

Jefferson or President Theodore Roosevelt. He did not carry on a voluminous and polemical correspondence with American and foreign intellectuals as these men had done, even when they were in the White House.
And Kennedy himself realized that his "little promotions" did not help
young and struggling artists and writers in the direct and material way
the New Deal works projects had done.

But never before Kennedy's time had the White House paid so much
personal and social attention to the nation's writers, artists, musicians,
scientists, and scholars. At first some of the public was inclined to take a
snidely skeptical view of all this. Was not this celebrity-hunting, high-
brow name-dropping, a further drive to presidential glamour? The recipients of these attentions did not think so. Only William Faulkner, in
bad-tempered petulance, rebuffed the President. For the rest, a chat with
the President or an invitation to an event in the White House was an
occasion of a lifetime, and these felt that Kennedy was not merely honor-
ing them but the creative work they represented. As Richard Rovere has
pointed out, Kennedy was tremendously concerned that the American
society become a good, even a brilliant, civilization. He thought of himself as a promoter, an impresario, of excellence in every phase of American life, and he hoped that future Presidents would emulate him in this.[9]

To latter twentieth-century Americans, Kennedy will be a kind of beau
ideal reflecting what they consider admirable in the politician—a shun-
ning of corniness and hokum, an accent on youth and wealth, the glamorous videographic personality favored by Hollywood and TV, a contrived
casualness in dress and manner, the sophistication and urbanity of the Ivy
League, direct and clear speech sprinkled with wit, an avoidance of doctrine and dogma, a pragmatism just emotionally enough involved to be
effective, the capacity for using expertise and Madison Avenue techniques, the ability to create and sustain an "image." In these, most of
them externals, Kennedy will have many imitators.

The Kennedy élan will not be easy to imitate. Even more difficult of
imitation will be the Kennedy mind—rational and balanced thinking,
objectivity, the ability to see all around a question, resilience, elusiveness,
the capacity for keeping judgment in suspense, a detachment reaching to
one's self and one's own image, an avoidance of absolute commitment
combined with genuine intellectual involvement, a general critical intel-

9. See Richard Rovere, "Letter from Washington," *The New Yorker*, November 30,
1963.

ligence brought to bear on the findings of the specialists. The Kennedy magic lies in its combination of the various elements: the externals, the verve with which the externals were carried off, and the cast of mind.

There is still another Kennedy intangible, perhaps the most important, one which belongs to the nonrational. Kennedy is becoming a folk hero, a subject of myth and legend, one of those few in history who capture the poetic imagination and affection of the masses. Solid achievement may have something to do with arriving at such a place in history, but very often it has little or nothing to do with it. Indeed, the titans who have wrought most mightily and in the end been felled by tragedy inspire awe and reverence more frequently than they do folk affection. They are too mature, their lives too devoid of colorful gallantries and foibles, their achievements too overwhelming for the average man to identify himself with such figures. To this class belong Caesar, William the Silent, Lincoln. Increasingly Lincoln has become a father image and "the martyred Christ of the democratic passion play."

The folk hero in the affectionate, indulgent sense is one who leaves behind him an over-all impression of élan, style, beauty, grace, gaiety, gallantry, bold and light-hearted adventure, valor—mingled in one way or another with the frail, the fey, the heedless, the mystic, the tragic. This is the romantic tradition, the tradition of Achilles, David, Alcibiades (despite his damaged soul), Arthur, Roland, Abelard, Richard the Lion-Hearted, St. Francis, Bayard, Raleigh, Henry of Navarre, Gustavus Adolphus, Byron. Alexander the Great is often put in this tradition, but his exploits were so dazzling, so epoch-making, that he became more a god than a hero.

Kennedy's death has in it the touch of religious epic, of man pitted against fate. Here surely was one favored by the gods, one possessed of power, wealth, youth, the aura of manly war heroism, zest for living, personal charm and beauty, glamour, imagination, keen insight, intelligence, immense popularity, the adoring love of family and friends. Great achievements were to his credit, and even greater ones seemed in store. Then in the fullness of his strength, he was cut down in a flash. History has no more dramatic demonstration of the everlasting insecurity of the human condition.[10]

10. Although the folk think of Kennedy as a child of fortune, actually he suffered much physically. He seems never to have been robust; the state of his health often concerned his father; in my own brief Kennedy files are letters from the elder Kennedy which speak of Jack's poor health in periods prior to World War II. Following

Was Kennedy himself a romantic? In some ways, mostly in appearance and manner. There are photographs of him, for instance several public ones taken in Tampa five days before his assassination, which reveal him in a kind of narcissistic euphoria. (Those who understand how wondrously flexible human nature can be will see nothing damaging in this.) James Reston once observed that the effect Kennedy had on women voters was "almost naughty." In his personal relations—and this is a matter not of appearance but of substance—Kennedy had an outgoing freshness and (there is no other term for it) a sweetness of temper. But basically Kennedy was not a romantic. He was a rationalist with a critical intelligence, a realist who knew the hard and subtle uses of power.

However, one need not be a romantic to become a romantic hero of history. Many romantics miss it—sometimes for a variety of reasons just barely miss it: Bolívar, Garibaldi, Gambetta, Jaurès, Michael Collins. In modern times romantic heroes have become rare. Kennedy is the first in this tradition in a long time, and he is the only American in its top echelon. Strange that he should have come out of the America of the machine and mass production. Or is it? People in our prosaic age, particularly young Americans, were yearning for a romantic hero, as the James Dean cult among our youth revealed. Now they have an authentic one.

---

his service in the war, Jack was plagued with malaria and his serious back injury. Even after the successful operation on his back, it would appear that he was rarely free from pain. When this comes to be realized, Kennedy's pace as President will appear even more gallant.

ᕲᕲ The Conservative Myth

There is a considerable body of literature that either explicitly or implicitly likens the so-called conservative movement in the America of our time to neo-fascist and crypto-fascist movements abroad. The following article uses the historical method to analyze today's self-styled American Conservatism and concludes that essentially it is indigenously American, that its prevailing attitudes represent an exaggeration and a distortion of values reaching far back into American history and deeply rooted in the American folk mind; but that, since its approach is anachronistic rather than relevant, romantic rather than realistic, it cannot come to grips with today's actual problems and must continue to baffle and frustrate its own followers. When intellectuals drawn to this movement attempt to engraft on a simplistic Lockeanism and a spread-eagle Americanism a Neo-Thomism and a Burkeanism imported from an authentic European conservatism, they confuse rather than clarify and give the movement a transparently fraudulent appearance.

■ DURING THE PAST YEAR, I made a number of college and university lectures in various parts of the country and frequently ran into speakers representing the so-called conservative movement, a few of them connected with *National Review*. At times I participated in forums and symposia along with "conservatives," and occasionally these developed into "debates." Since I live in an area of the country where Goldwaterism and the ultra groups to the right of Goldwaterism have much strength, I had a fairly accurate notion of this "new conservatism," but these platform experiences forced me to clarify my thinking and to examine anew the nature of the various groups which compose this movement and the internal inconsistencies and contradictions which beset it.

Frankly, I was surprised at the large number of college and university appearances the big names of the movement were making. The lecture schedules of some resembled the crowded itineraries of political candidates in the heat of campaigns; obviously "conservatism" was becoming a

The Antioch Review, XXVII (Fall 1967).

lucrative business. Usually these lecturers had contacts with "conservative" student leaders and groups and spent considerable time with them. In my own platform confrontations with "conservative" speakers, I often had the feeling that my opponents were more interested in indoctrinating and making political converts than in throwing intellectual light on public problems and trends, that they were primarily intent on making "points" after the fashion of high school debaters.

What surprised me, above all else, was that these "conservatives" were actually succeeding in appropriating the conservative label. Not once did any member of a college or university audience ever challenge the honesty or accuracy of these lecturers when they applied that label to themselves and their movement. This could happen only in a country without a genuine conservative tradition.

II

A melange of assorted groups make up the so-called conservative movement.

By far the most important is composed of those we shall call (for lack of a more precise term) the Goldwaterites. The basic political strength of the movement comes from these Goldwaterites, and without them it would be reduced to a mere fringe agitation.

The Goldwaterites are recruited from the old McCarthy following in the big cities (in large measure Irish-Catholic and passionately anti-Communist); from many Protestant fundamentalists; from strike-it-rich oil producers, particularly in the Southwest; from wealthy farmers (often connected with the American Farm Bureau) who are not dependent on the federal farm program for survival and are in a position to indulge the nostalgic individualism of the husbandman; from those involved in the more "capitalistic" types of commodity production, such as cattlemen, citrus-growers, and producers of specialty fruits and vegetables, all of whom are not tied as closely to a federal farm program as are small wheat-, corn-, and cotton-growers. The most powerful support for Goldwaterism comes generally from the new-rich entrepreneurial, managerial, and professional classes in the provincial towns and cities and their neighboring suburbs. Real estate brokers, insurance agents, building contractors, and medical doctors seem to be among those most frequently drawn to Goldwaterism. The most vocal Goldwaterites are found among those members of civic clubs, chambers of commerce, and country clubs who are strenuously "on the make," the promoter and booster types.

The heaviest concentration of Goldwaterite strength is in the "sunshine states" stretching from South Carolina to Los Angeles; and Goldwaterism is most spectacularly expressed in Florida, Texas, Arizona, and southern California. A number of factors account for this Goldwaterite concentration in the southernmost tier of states: phenomenally fast-growing populations, much of them transplanted; a new and rapidly developing industrialism; a boom psychology and transitional opportunities for real estate promotions and other speculations; the ever-increasing number of migrated and retired military personnel, *rentier* classes, and old folks of modest means fearful of both taxation and inflation; the importance of cattlemen and producers of "fancy" fruits and vegetables; the prevalence of racist attitudes in populations largely of Southern origin.

A number of lawyers and publicists are drawn to Goldwaterism by genuine fears of "Federalia" and by the intellectual attraction of the tradition of Quiddery in American politics and constitutional law. Even at the time of the fight over the adoption of the United States Constitution (1787–1789) and since that time, the Quids (an unjustifiably derogatory term meaning "quibblers," first applied to John Randolph's factional opposition to President Jefferson) have feared that the Constitution contains within itself the seeds of an infinite expansion of the federal government at the expense of the states, and they have always weighted the federal system in favor of the states, and sometimes even in the direction of a *confederative* rather than a federal interpretation.

Examples of Quiddery and of its tenacious persistence are legion: Patrick Henry and George Mason resisting the ratification of the Constitution itself; John Taylor of Caroline and John Randolph of Roanoke suspicious of Presidents Jefferson and Madison for their "centralizing" tendencies; John Tyler, President by a series of accidents, vetoing one national measure after another on strict states-rights grounds; Calhoun spinning his complicated constitutional theories to check federal power; Alexander H. Stephens denouncing the war measures of the Confederacy and Clement L. Vallandigham attempting to sabotage the war efforts of the Lincoln administration out of their strict-constructionist principles; the United States Supreme Court, in the early years of this century, interdicting labor and welfare legislation on doctrinaire laissez faire grounds.

Since 1789, those holding Quiddian views have on most issues been a distinct minority in the nation, but the vitality of Quiddery is remarkable. It reflects itself today in the editorials and columns of J. J. Kilpatrick, brilliant Richmond editor, and in his numerous imitators. How frequently the Goldwaterites quote Patrick Henry in pamphlets and speeches and insinuate quotations from him in college anthologies, with never a hint that these quotations are from Henry's speeches in opposition to adoption of the Constitution itself!

Those in search of a home-grown historical and intellectual rationale for the contemporary "conservative" movement will find it in Quiddery, and not in *National Review*'s hothouse Burkean posturings.

To the right of the Goldwaterites are the ultra groups—the John Birch Society, the National Indignation Convention, Project Alert, the Christian Crusaders, the Anti-Communist Christian Crusaders, the Minutemen, the Ku Klux Klan, and others. These are more emotional and exuberant than the Goldwaterites, more fundamentalist and intolerant, more patrioteering and more doctrinaire anti-Communist, and they are often impregnated with vigilanteism and more overt racism. The Goldwaterites welcome some of these ultra groups, outwardly repudiate some of them, co-operate willy-nilly with most of them—and usually win the votes of the members of all of them at election times.

The Goldwaterites, and, yes, most of the ultra groups to the right of them, exemplify indigenous American folk attitudes. They are rooted in America's broad liberal tradition, in what Louis Hartz calls Lockeanism. Ideologically, the Goldwaterites and the ultra groups to the right of them distort the free enterprise aspect of Lockeanism, give it a lip-service literal laissez faire interpretation which does not square with the way it actually operated throughout our history, and their Quiddery is merely an exaggerated Lockeanism applied to politics and constitutional law. (Where their own pocketbooks are concerned, however, the Goldwaterites are rarely above calling on the federal government for help.) Despite certain racist and vigilante overtones among some of the "conservative" groups, it is misleading to apply such terms as "fascist," "neo-fascist," or "crypto-fascist" to their basic economic, social, and political values, for these are emphatically home-grown American and not European. Even the racist and vigilante proclivities have their origins deep in American history—in the ugly treatment of Indians and Negroes through more than three centuries, in frontier customs, and in our puritan penchant for meddling with the other fellow's thought and conduct.

In order to give "tone" to the "conservative" movement, *National Review* attempts to link the Goldwaterites to the Neo-Thomists, the Burkeans, and the Southern Agrarians. Now, the number of Neo-Thomist,

Burkean, and Southern Agrarian intellectuals in America is small; only a portion of these are willing to lend themselves to the movement; and the ones who do represent an alien element in the movement's mainstream. Their thought is not reflected in American folk attitudes and traditions. They bring little popular support and precious few votes. The attempt to assimilate the folk values of the Goldwaterites to Neo-Thomist and Burkean thought produces confusion and exposes the movement to a host of theoretical inconsistencies and contradictions.

The influence of the Neo-Thomists, who update and intellectualize traditional Catholic thought, is virtually nil in the practical life of America. Even well-educated Catholic laymen in this country usually know little about the great writers from Aquinas to Maritain who have contributed to Catholicism's humane traditions. As the Catholic immigrants increasingly assimilated themselves to American life, they unconsciously took on, in their secular conduct and activities, America's dominant Lockean, individualistic, and pragmatic values.

As to the Burkeans, Russell Kirk, *National Review*'s best known Burkean contributor, finds in Burke a body of first principles which includes belief that a divine intent rules society; affection for the mystery of traditional life; conviction that society requires orders, classes, status, social gradations, and hierarchy; persuasion that property and freedom are inseparably connected; faith in prescription; recognition that change and reform are not identical. This, of course, would do for a description of traditional conservatism in Europe, where societies emerged from medieval and feudal conditions; but it is a far cry from the American tradition, and it is utterly inconsistent with the "conservatism" of American business, with its restless dynamism, its passion for technological change, its faith in progress, its skillful catering to mass consumer demands, its constant search for "new blood" and promising young "comers."

The literary Southern Agrarians are basically Burkeans, for their thinking goes back to the Southern Conservative Renaissance of the 1850s, which portrayed the Old South as an integrated class-structured society. But this, too, is unrealistic, for the Old South was essentially a semi-frontier society; its white class lines were fluid; its so-called aristocracy was parvenu, mostly drawn from yeomen farmers; and the wealth of this "aristocracy" was as much derived from aggressive promotions and shrewd land speculations as was that of the successful men of the North and West.

III

The Goldwaterites push to the point of caricature those attributes and values we think of as peculiarly American. Their exaggeration of Lockeanism is unrealistic and in this age anachronistic. They convert aggressive competition into a sentimental and romantic Horatio Algerism. They pay enormous lip service to the old puritan attitudes at a time these have mellowed. In general, they are even less sensitive to Western civilization's tradition of secular humanism—its respect for classical and Renaissance learning, its body of theoretical thought, its aesthetic satisfactions, its spirit of *joie de vivre*, its urbanity in living—than are other groups of Americans.

In America, the absence of feudalism, of a class-structured society, and of corporative institutions inherited from medieval times, together with a small population and an abundance of land and other resources, produced a natural Lockeanism, a widely free market, an individually competitive society—indeed, a "monolithic liberalism." Americans prized liberty and democracy as instruments to make the economic game an open one, to guarantee access and fluidity, to provide equality of opportunity in the pursuit of wealth, to stimulate individuals to "get ahead." Goldwaterites concern themselves, above all else, with American free enterprise, its driving acquisitiveness, its intense urges to business success. Who can deny that these have been central to American history and life?

Nobody has ever better described the speculative and promotional drives of Americans, their restless dynamism, their relentless Faustianism, than Anthony Trollope, as perceptive an observer of the American scene as his mother, Frances, but more circumspect; and Trollope never descended to the excessive caricaturing found in Charles Dickens's *American Notes*. Trollope, traveling in America in the mid-nineteenth century, discovered that Americans regarded even shady promotions, "gold-brick" stock companies, and wildcat banks as playing a "legitimate" part in tapping savings, creating credit, and opening up their country; that most Americans gambled for the unearned increments of rising land values; that Americans were as enamored of the sheer game of money-making, in which some cheating and swindling was expected, as of the money itself; and over and again he heard Americans say in effect: "In a new country, sir, it behooves a man to be smart." A buoyant, optimistic enterprise was the prevailing mood of society. "In no country has pecuniary ruin been so common as in the United States; but then in no country has pecuniary ruin been so little ruinous."

Trollope, himself the product of a Burkean society, was astonished at the bustle of enterprise in America's non-Burkean one.

In the principal business streets of all these towns one sees vast buildings. They are usually called blocks. . . . I have had occasion in various towns to mount the stairs within these blocks, and have generally found some portion of them vacant—have sometimes found the greater portion of them vacant. Men build on an enormous scale, three times, ten times as much as is wanted. They only measure size on an increase of what men have built before. Monroe P. Jones, the speculator, is very probably ruined, and then begins the world again, nothing daunted. But Jones' block remains, and gives the city in its aggregate a certain amount of wealth. Or the block becomes at once of service and finds tenants. In which case Jones probably sells it and immediately builds two others twice as big. . . . He is greedy of dollars with a terrible covetousness; but he is greedy that he may speculate more widely. He would sooner have built Jones' tenth block, with the prospect of completing a twentieth, than settle himself down at rest for life as the owner of a Chatworth or a Woburn.[1]

Certainly this compulsive acquisitiveness, this unquenchable spirit of business enterprise, this perennial optimism, these passionate urges to promotion, speculation, material wealth, bigness, and "progress" constituted a bundle of basic and indigenous American cultural values. Indeed, these cognate values frequently came to be equated with Americanism itself.

Trollope confined his observations to the East and the states we now call Midwestern. But the South, too, was immersed in business enterprise and in promotions and speculations. Its colonial gentry had risen largely on the strength of shrewd speculations in land, slaves, and commodities. Its post-Revolutionary "aristocracy" was even more parvenu and given still more to economic plunging. In a number of ways, the planters, mostly sprung from the ranks of yeoman farmers, were tied closely to the Charleston and New Orleans merchants, brokers, and bankers, who were also largely "self-made" men. The expansion of the cotton kingdom into Alabama, Mississippi, Louisiana, Arkansas, and Texas (in the 1820s, 1830s, 1840s, and 1850s) widened still further the scope for men "on the make" to exploit virgin soil ruthlessly and gamble in land, slaves, and commodities.

Mack B. Swearingen has described how, in the Mississippi of the 1820s

1. *North America* (New York: Harper and Brothers, 1862), pp. 118–119.

and 1830s, a poor farmer often leaped from log cabin to mansion in the space of a few years. Swearingen found the way of life of these new cotton nabobs "savage"; they reminded him "more of Mongolian princes than of American citizens." By the 1850s the planters of the new cotton kingdom had become the "swell-heads" and the "cotton snobs." That indefatigable traveler and reporter, Frederick L. Olmsted, observed that the planters of Louisiana and Mississippi, along the Mississippi River, were for the most part "the immoral, vulgar, and ignorant newly rich." Their conversations ran almost exclusively to the prices of land, slaves, and cotton; their amusements were largely drinking, gambling, and horse-racing. The pretentiousness of these newly rich reached ludicrous proportions in the "villa district" which radiated from Natchez.

From whence had most of the founders of these "aristocratic families" come? From the farmer and overseer classes. Did their sons manage to hang on? Usually, yes, if they did not meet premature deaths in drinking bouts, horse-racing accidents, duels, and brawlings. How about the grandsons? "Almost always they are fools, and soon go through with it." And who succeeded to the estates and the slaves? The mortgages were taken over by other planters, or by "nobodies" from the ranks of farmers, overseers, or city mercantile classes.

In this free-wheeling enterprise system of "the elegant Old South," it took drive, ruthlessness, and sharp speculative instincts to reach the top of the social heap; and in order to remain there, it took, in addition to these qualities, a lot of care, hard work, and luck.

Today, well past the twentieth-century mark, Americans have become a more decorous people, and many of these nineteenth-century drives and characteristics, so well described by Trollope and Olmsted, have been chastened. These still persist, however, as powerful cultural residues in all parts of the country, most of all in fast-growing boom areas of the South and West, the very areas where Goldwaterism is strong.

But the Goldwater free-enterprisers differ in several important ways from their nineteenth-century prototypes. For one thing, today's most flourishing boom areas are dependent on defense, space, atomics, electronics, and supersonic transport industries; and it is government that underwrites most of their development costs and buys the bulk of their output. The "self-made" prosperity of local realtors, contractors, insurance agents, automobile distributors, and other entrepreneurs in these areas has in fact been generated by government. For another thing, even the entrepreneurs little connected with the government-stimulated industries (the cattlemen and citrus-growers, for instance), while "free-enterprisers" with respect to their profits, look to government to "socialize" their losses. For still another thing, today's free-enterprisers are a tamer breed than those of the nineteenth century; they are far more afraid of losing their new wealth than were their nineteenth-century predecessors.

IV

A purely free-market economy has never existed in even America. What has been the role of government in America's predominantly Lockean economy? David M. Potter has well expressed the continuing part played by government:

If we are to appreciate the links with the past, we must recognize that laissez faire was not the unique principle of policy in our eighteenth- and nineteenth-century development but that one of the key principles was certainly the constant endeavor of government to make the economic abundance of the nation accessible to the public. The tactics by which this was done changed as the form of abundance itself changed, but the basic purpose—to keep our population in contact with the sources of wealth—has remained steadily in the ascendant throughout our history.[2]

During the nineteenth century, this governmental endeavor to put our population in contact with the sources of wealth was manifested in a number of ways; and all political parties and groups, within the framework of America's inclusive Lockeanism, contributed to enlarging individual access to material abundance. The Jeffersonians and Jacksonians were in the forefront in widening access to the public lands, although in the end it was the Republicans and not the Democrats who pushed through the climactic homestead law. On the other hand, it was the Federalists, Whigs, and Republicans who did more to promote domestic markets, trade, and industry by sponsoring public support of schools (to prepare individuals to function in a commercial-industrial society), by advocating the chartering of banks and credit agencies, by agitating for tariff subsidies, and by fighting for government aid to internal improvements, river and lake navigation, canals, and railroads; although it was the Northern Democrats in the days of Jackson and Van Buren who took the lead in smashing what was known as monopoly mercantilist capitalism and establishing liberal capitalism.

Despite the humanitarian and democratic philosophy of Jefferson,

2. *People of Plenty: Economic Abundance and the American Character* (Chicago: University of Chicago Press, Phoenix Books, 1954), p. 123.

many Europeans, drawing analogies from experiences in Europe, where it was the commercial classes who first broke out of the feudal mold, insisted on thinking of the Federalists and the Whigs as the dynamic forces in American politics, as the true revolutionaries. Hamilton's system of a perpetual public debt and a national bank was likened to that of Charles Montagu and the English Whig Junto after the Revolution of 1688. And when in the 1830s and 1840s the American Whigs gave up the aristocratic social pretensions of their Federalist predecessors, abandoned monopoly mercantilist capitalism for liberal capitalism, and consummated a marriage of democracy and capitalism by proclaiming the from-rags-to-riches Horatio Alger myth, an even larger number of Europeans were convinced that it was the Whigs and their Republican successors, and not the agrarian Democrats, who were in the forefront of the battle to widen access to America's fabulous natural resources through commercial and industrial revolutions.

Later on, the writings of Karl Marx on American politics reflected the views not only of Marxists but of Europeans generally about American political parties. Marx regarded the Democrats as the representatives of an agrarian feudalism, the Republicans as the exponents of an advancing industrialism. He saw the Republican crusade against slavery as only a part of the larger American bourgeois revolution, which required a free labor system.

The Whigs and Republicans themselves believed they were the better exponents of "dynamic American progress" than were the Democrats. As late as the turn of the century, John Hay often spoke contemptuously of the Democratic party as "a fortuitous concourse of unrelated prejudices," meaning in plain English that it was the party of transplanted Irish-Catholic peasants and of Southern fundamentalist Protestant yahoos. Theodore Roosevelt often disdained the Democratic party as the party of "rural Toryism."

It should be noted that during the nineteenth century, government was chiefly concerned with putting the entrepreneurial classes into contact with the sources of American wealth, in the belief that if these classes were prosperous, economic well-being would permeate all other classes in the society. But in the twentieth century, an advanced industrialism raised problems requiring a larger government intervention, among other things measures to insure a constant flow of purchasing power to allow a consumption of the avalanche of goods being produced by that time. Government then turned to measures designed to give the nonentrepreneurial classes a more direct access to American plenty. All of this was accomplished not by repudiating the Lockean heritage but by specific and pragmatic "problem-solving" within the framework of a dominantly private-capital entrepreneurial system. Nor were the entrepreneurs in fact left out; indeed, in addition to profiting indirectly from the Keynesian policies of keeping the economy on an even keel, many became the chief direct beneficiaries of federal government policies—real estate brokers and contractors under the federal housing programs, owners of federally chartered building and loan associations, private insurance companies acting as the agents of the federal government under Medicare, and so forth.

As the twentieth century advanced, it was the Democratic party, responding to the new problems of an advanced industrial society and spearheading those measures designed to put an even larger number of Americans in direct contact with American abundance, which now laid claim to being the better champion of American progress.

In short, in the nineteenth century, government measures were largely concerned with putting the entrepreneurs in contact with the sources of wealth; but in the twentieth century, government measures are designed not only to keep the entrepreneurs in contact with the sources of wealth but also to give the nonentrepreneurial classes larger access to America's economic abundance. However, our new "conservatives" continue to think of free enterprise, including government aid to business, as an entrepreneurial preserve; they regard the new governmental intervention, to widen access of the nonentrepreneurial classes to American plenty, as a betrayal of free enterprise, as "socialism."

V

For over a decade, I have been observing "conservative" and right-wing gatherings in the provinces. When a literary Southern Agrarian or a Burkean is the speaker before a general audience, he is a "dud," although he may evoke a genteel response when speaking to a group of fashionable women or to a provincial college audience.

It is a different story, however, when an Eddie Rickenbacker, a Clarence Manion, or a Billy Hargis addresses a popular meeting. The audience is carried away by a stem-winding oratory which claims that true liberty was born into the world with the American Revolution and the defeat of the Tories, praises the old-time spartan and puritan virtues, extols rugged individualism, and equates Horatio Algerism with Americanism. Many of these popular orators invoke the name of Jefferson in defense of local management of affairs, states' rights, and little government. In acknowledging their debt to Jefferson, they are (consciously or unconsciously) acknowledging their debt to Locke.

Who can doubt that this represents a genuine American gospel, an indigenous American folklore? Much of it is right out of the McGuffey Readers.

To thoughtful observers, though, there is poignancy in this oratory, and in the response to it; for while it reflects an authentic American tradition, it oversimplifies and distorts that tradition, appeals to a Lockeanism which never in fact existed in the simple way portrayed, reduces Jefferson to a mere doctrinaire, and obscures rather than clarifies the problems Americans face today.

National Review has the hardihood to attempt to infuse Goldwaterism with humanism, and to reconcile Goldwaterism's exaggerated free-enterprise doctrines with Burkeanism.

Editor William F. Buckley is apparently a Tory humanist with a respect for the classics, *belles lettres*, personal privacy (for those who can afford it), and an aristocratic liberty which extends to eccentricity (for those with the taste to carry it off). He questions technocratic values and seems genuinely concerned about preserving old humanist ones. But he cuts an ironical figure when he represents classical and humanist values as those of the Goldwaterites or attempts to sell such values to them. For the most vocal of the Goldwaterites are hell-bent for the technocratic society. (Goldwater himself is a tinkerer in the tradition of *Popular Mechanics*.) These Goldwaterites are above all "go-getters," materialists, conspicuous consumers, and status-climbers much in evidence in the country clubs and along the posh cocktail circuits, many of them characters right out of a John O'Hara novel. They welcome technological innovations as prospects for speculations and profits, even though the ultimate result must be to move us more relentlessly to a weblike technocratic society. If they could, they would reduce even the air, the rain, and the sunshine to gadgeteering profit-making. It would seem that nature intended Buckley to essay the role of another Mencken, not that of a missionary to the primitives of "the Sahara of the Bozart" and the Babbits of Main Street and Suburbia.

The clash between *National Review*'s Goldwaterite contributors and its Burkean ones (like Kirk) is also too glaring to disguise, and the claims that the "conservative" movement is rooted in Burke emerge as pure pose.

The inherent incompatibility between Burkeanism and American "conservatism," indeed between Burkeanism and the American tradition, was trenchantly pointed out in a letter to *National Review* (August 13, 1963) from a genuine Burkean, J. W. Daly, a member of the history faculty of Christ the King College, London, Ontario. Buckley published the letter in full—out of grace, or with a puckish chuckle that recognized (and perhaps sympathized with) the truth of Daly's insight? Here is the letter:

> Your editorial, "The Vote in the South," is just too much; it is a beautiful example of the American conservative dilemma. How can any American decry "the popular mania for the franchise" with a straight face? Your Revolution, the ruthless act of dedicated terrorists, had to justify itself on a similar mania— "the sovereignty of the people." Given such abstract doctrine, which Burke, like any intelligent conservative, utterly scorned, how can such equally abstract dogma as the universal franchise be denied? If "the people" are sovereign, how can one group of them presume to have a voice when another does not?
>
> Quit trying to hide behind the Federalists. They came *after* the Declaration of Independence, and would have been impossible without it. They, and you, cannot shrug off that bitter legacy.
>
> American ideology is based on a lie. I do not dispute its value; my countrymen may perish for lack of such a convenient fiction. I merely point out that no American can place himself outside that lie. The tragedy of American conservatism lies in its inescapable defense of a Liberal dogma.

Even if one believes that the concept of the American society as a "monolithic liberalism" is an oversimplification of the facts, he must certainly agree that the American society has never been—even remotely—a Burkean society. (Hartz points out that the Americans are "Burkean" only in the sense that they respect their Lockean tradition; that in America Burke means Locke.) From colonial times to the present, European observers, themselves members of traditionally Burkean societies, have said an infinite number of things about the American society, but none has ever described it as Burkean.

The intellectuals among America's "new conservatives" are ideologically at sea. Their pure Lockeanism would have been a distortion even a century ago; today it is an anachronism. Their Burkeanism is romantic and literary; it has no footing in the American tradition. And when these "conservatives" attempt to embrace simultaneously both a Lockean caricature and a Burkean irrelevancy, they make themselves ridiculous, and they give their movement a transparently opportunistic, even fraudulent, character.

VI

The world view held by this "conservative" movement is out of focus. Our "conservatives" have never accepted the fact that the Communist revolutions in Russia and China are there to stay. At the same time, they cling to the myth of a monolithic international Communist conspiracy relentlessly moving to take over the world. Non-Communist leftist and neutralist governments are seen as stooges and fifth columns for the Communist conspirators. Our "conservatives" reject not only a *détente* but even "the snare" of peaceful coexistence with any Communist power. Hence, instead of standing for measures which, while safeguarding national security, seek to stabilize the world, our "conservatives" plump for American unilateralism abroad and needlessly provocative and perilous risks threatening ultimate nuclear holocaust. Because of their extravagant fear of Communism, our "conservatives" scuttle their own arguments for less government spending and less government in general by sponsoring ever-larger military appropriations and an ever-larger military establishment (beyond what liberals, moderates, and authentic conservatives deem necessary to the national security) and by advocating measures which would move us in the direction of a Leviathan police state, a garrison society, and a brass-hat socialism.

International wars have been the great change-makers of history, both at home and abroad. Such wars invariably unleash unforseen forces which drastically transform the old prewar world. A third world war, inevitably a nuclear one, would revolutionize all societies and alter our world beyond recognition. The instinct of genuine conservatives is to support measures for national security but at the same time to take advantage of any conditions and opportunities that give promise of accommodation, stability, and peace in international affairs. For instance, true conservatives welcome the shattering rifts within the old Communist bloc as probable harbingers of a less dangerous world; but America's ersatz conservatives either insist that these are Communist hoaxes or resent them as belying their rigid view of the world and robbing them of their chief propaganda stock-in-trade, that is, the myth of the monolithic international Communist conspiracy.

To those who took the American folklore literally, the last few decades in world affairs have been a time of repeated and painful shock. This folklore held, among other things, that the poor and oppressed everywhere looked to America, and to no other land, as the country of opportunity and the good life; that the Americans were the undisputed leaders in modern technology; that America enjoyed a natural immunity from foreign threats to its national security. But now other ways of life have arisen to compete with the American way for the affections of the poor and underprivileged. The Russians, until a few decades ago a backward peasant people, have speedily industrialized themselves, and they have challenged Americans with technological innovations, particularly in the field of outer space. American national security and even survival no longer come free; Americans must work and sacrifice for these boons every day in the grubby arena of world politics, and nuclear weapons have made Americans as vulnerable to destruction by foreign enemies as other peoples have always been vulnerable to invasion and devastation.

This loss of innocence, this vanishing of the traditional certainties, and this collapse of the old immunities have been a particularly traumatic experience to the less informed, the less historically insightful, and the less resilient of our people, compelling them to seek compensations in super-patriotism, scapegoats, and the conspiratorial view of history.

Thus both at home and abroad, our "conservatives" have responded to contemporary challenges by distorting American folk traditions—by an extravagant Lockeanism in domestic affairs and a spread-eagle Americanism in foreign affairs.

VII

Despite the smashing defeat of Goldwater himself in 1964, the strength of Goldwaterism and its allies is far from spent. The Goldwaterites continue to win local and even statewide elections in some of the Southern and Western states. They will be a force to be reckoned with in the Republican national convention of 1968. A George Wallace third-party candidacy that year will also be formidable, not only in Southern states but in a few Western states as well; and it may have a decisive influence on the national outcome.

What do prosperous Goldwaterites want in politics and in life? (And it is the prosperous ones who call the tune in "conservative" politics.) They resent increased taxation for schools, colleges, universities, libraries, hospitals, welfare payments, old-age assistance, and control of air and water pollution. They prefer instead special exemptions and inducements to attract new industries, and lower taxes to allow well-to-do citizens to make additional private investments and to indulge a third automobile, a fourth bathroom, a private swimming pool, a more elegant country club, better golf courses, and a winning football team at the state university.

How do the prosperous Goldwaterites manage to attract large popular support? By their promises of less government spending and lower taxes; by their inveighing against crime, criminals, juvenile delinquency, "obscene" literature, lawlessness, riots and street disorders; by their racist insinuations; by their anti-Communism; by their patrioteering. At election times they attract, either overtly or covertly, the ultra groups to the right of them. The "conservative" emphasis on the myth of an international Communist monolith steadily taking over the world has helped the Goldwaterites win elections, for many Amercians who do not embrace any of the other facets of the "new conservatism" continue to believe this myth. Its persistence in the popular mind has doubtless contributed to the continuing inability of American foreign policy to adjust adequately to the new polycentrism and fluidity in the world today.

In the South, Goldwaterism has two political factions, one composed of dissident Democrats and the other of nonmoderate Republicans. The former are more racist, more rural, more folksy, more church-going, and more old-fashioned. The latter are less racist, more urban, more economic-minded, more worldly, "smoother," and more intrigued by professional public relations, the fashioning of "images," and the "expert" manipulation of stereotypes and symbols. Both factions often come together in electoral alliances, and they co-operate across formal party lines in state legislatures.

These are the attitudes, the issues, and the combinations which largely account for the popularity of a Murphy and a Reagan in California, a Goldwater in Arizona, a Tower in Texas, a Williams in Mississippi, a Wallace in Alabama, a Kirk in Florida, a Maddox in Georgia, a Thurmond in South Carolina.

VIII

There are, of course, Americans with profoundly conservative instincts. Like the Goldwaterites, most of these true conservatives are rooted in America's broad liberal tradition. Clinton Rossiter has denominated Walter Lippmann the nation's leading conservative, and Lippmann's *Good Society* is one of the most eloquent defenses of the market economy written in this century. But the Lippmannites have a deeper historical and social understanding than have the Goldwaterites; they are more concerned with infusing economic values with humanist ones; they

185

are more willing to come to realistic terms with the drastically altered conditions of our time; they are more aware of the profound and permanent changes wrought by the scientific and technological revolutions, and they seek adjustments without which even more of the past would be swept away by mounting tensions, repeated crises, and eventually angry eruptions.

Lippmann reveals a genuine lament for the past not found in the facile "conservatives." He is more keenly conscious of how difficult it will be to preserve traditional values in our technocratic age. He has eloquently observed that to the authentic conservative, as distinguished from the romantic one and the deluded reactionary, the poignant question, not yet answered, is how, with the ancestral order dissolved and the ancient religious certainties corroded by science, contemporary man can find "meaning which binds his experience and engages his faculties and his passions." Again: "The modern conservative has to work in unprecedented ways for an undefined future. He has to create the new forms in which the enduring truths and values can be carried on in a world which is being rapidly transformed."

Just as the Goldwaterites are successfully appropriating the term "conservative" for themselves, so also they are succeeding fairly well (at least in the provinces) in identifying Lippmann with "the Liberal Establishment." To repeat, this could happen only in a country without a genuine conservative tradition.

Robert A. Nisbet, another true conservative, believes that what most disturbs those people deeply troubled about the dominant social and political trends of our time is the growth of big and centralized government and the increasing acceptance of the unitary political rationalism of Plato, Hobbes, and Rousseau, which sees private institutions and private powers "as distractive of common purpose, inimical to common welfare, and potentially subversive to common government." Nisbet suggests that a philosophy to bind together in intellectual integrity most people today who are alarmed at current trends may be found in a nineteenth-century variant of seventeenth-century and eighteenth-century Lockeanism—in the tradition of Lamennais, Tocqueville, Mill, Acton, and Max Weber, who held the existence of autonomous private associations and private powers to be necessary structures of intermediate authority in society, buffers protecting individual freedom, and indispensable checks on the centralization of the state and on an atomized mass society.

Nisbet's concept of liberty would combine the Lockean (and Quiddian) emphasis on a formal bill of rights and contractual constitutional restraints with equal attention to the vitally libertarian part played by diversity, pluralism, and variegated group organizations in society itself. If adopted, this would again demonstrate the all-inclusive nature of America's liberal tradition, for it would underscore the fact that even genuine conservatives in America must depend for a unifying philosophy on a Lockeanism modified, supplemented, and made relevant by an emphasis on the active interplay of autonomous group organizations. This combination, however, would serve merely as a beginning for a viable conservative economic-political rationale in today's world; and it would still leave unsatisfied the need to fashion contemporary religious, moral, educational, literary, and aesthetic forms for expressing those "enduring truths and values" which the instinctive conservative prizes highly.

Today, Americans of truly conservative bent, with intellectual origins in America's broad and indigenous liberalism, stand apart from the Goldwaterites and the right-wing ultras, from the exotic Burkeans and Southern Agrarians, and from *National Review*. Their number is increasing; and they are searching, against great odds, for a body of authentically conservative principles applicable to our galloping technocratic society.

III

*Changing Cultural
Ethos and the
Technocratic Society*

¬ Hawthorne Discovers the English

One cannot understand the distinguishing characteristics of the present-day technocratic society without some "feel" for the cultural ethos of the nineteenth century which preceded it. The following article attempts to capture the flavor of the English society during the middle decades of the nineteenth century, when England was becoming industrialized but the people were still predominantly rural. Throughout the article there are numerous comparisons of the English and American societies of the time.

The most palpable feature of English life a century ago was the pervasive poverty of both the rural and city areas. But America, too, was a land of widespread poverty. In the vast and far-flung rural sections, most people still lived in cabins and shanties amidst semi-frontier conditions. There was more access to a rude abundance than there was in the English rural villages, but there were also more personal hardships. America, too, already had its city slums; one, the Five Points in New York, was internationally infamous, and Charles Dickens described its horrors as luridly as he did those of the English slums.

Another conspicuous attribute of English society a century ago was its traditional class system. But the America of the same time, despite the fiction to the contrary, was also a class-conscious land, although the class lines were less rigid and more fluid. Every county had its planter, prosperous-farmer, creditor, land-speculator, promoter, entrepreneurial gentry, many of them among the earliest settlers of the area or descendants of such settlers. In the East there was a mercantile elite which dominated affairs in State Street, Wall Street, and Chestnut Street; and when New York's men of affairs gathered at their Hone club and Kent club dinners, they conducted themselves in much the same fashion and revealed many of the same attitudes and prejudices that Hawthorne observed when he attended the grand ceremonial dinners of the mercantile "princes" of Liverpool and London. It is only when we compare conditions then with conditions now that we can realize how far both the English and American societies have moved in the direction of egalitarianism.

In general, both the English and American societies of a century ago were more religious than they are today; more romantic and sentimental; more bucolic; more provincial; less formally educated; outside the

The Yale Review, LIII (Spring 1964).

crafts, less occupationally specialized; less dependent on mechanical contrivances; personally more self-reliant; less rationalistic in goals and personal orientation; less sophisticated; more playful and "natural," despite an affection for public ceremonials and the formal and ritualistic behavior at upper-class functions; personally more inner-directed and more prejudice-directed; less deliberately and less consciously other-directed.

■ EVERY LOVER of books has a few unique favorites, little known to others but precious to him. Topmost among my own particular favorites are two volumes which contain Nathaniel Hawthorne's *English Note-Books*, his personal journal, 1853–57, when he was American consul at Liverpool, and *Our Old Home*, impressionistic essays he distilled from portions of his journal. *Our Old Home* was published in 1863, the *English Note-Books* posthumously in 1870.

These are among the least known of Hawthorne's works. It is strange that Hippolyte Taine's *Notes on England*, treating English society in the early 1860s, is still regarded as a minor classic, while Hawthorne's *Note-Books*, less analytical but superior in imagination, description, and those nuances which an Anglo-American like Hawthorne could feel but a Frenchman and more remote outsider like Taine could not, is almost forgotten. But *Our Old Home* and the *English Note-Books* are not merely valuable for the light they throw on England and Englishmen; in committing to writing these observations and his own daily activities, Hawthorne necessarily sharpened his perceptions of America and Americans and revealed himself much more intimately than he did in his fiction. Thus for insights into Palmerstonian England (and even the England of today), into mid–nineteenth-century America (and even the America of today), and into Hawthorne himself, these volumes are rarities.

Apart from these larger considerations, what has drawn me to these volumes again and again, always with fresh relish, is the remarkable personal rapport I share with the author in tastes and interests: a partiality for heavy food (Hawthorne lived in a more compatible century); the cheer of a good dinner followed by hours of lively table talk; the soothing comfort of rain and the exhilaration in the clean freshness which follows rain; an aversion for nature in the wild and a preference for nature subdued and verdant; an attraction to old cemeteries and the social and human-interest lore to be found on antique tombstones; a never-ending interest in historic spots and a willingness to drag one weary foot after another at the prospect of fresh exploration; a distaste for guides with rote spiels and for anything else that gets in the way of uninhibited historical imagination; a sense of sacrilege at shiny restorations (how Hawthorne would have writhed at the world's-fair atmosphere of present-day Williamsburg!); a perpetual delight in Gothic architecture and the snug little squares of old European cities; a passion for roaming the streets of large cities, particularly their old and planless parts, the haunts of the poor, the frowzy, the seamy, the sordid, and the grim, where the ambiguities of our human condition can be more readily (but not more truthfully) felt than elsewhere.

In order to provide some system to Hawthorne's observations, I have taken the hundreds of entries in his journal, each of which contains a miscellaneous number of impressions, and have distilled, classified, and combined these impressions, thus giving a measure of synthesis to Hawthorne's ideas and making it possible to compare his conclusions with Taine's and to test them against the perspective of time—all, it is hoped, without unduly sacrificing the poetry of Hawthorne's delightfully discursive work.

Hawthorne observed that the climate was the only attribute of their country the English never overvalued. Taine thought the English climate abominable and ascribed to it both the emphasis on exercise that made English women hearty Amazons and the heavy drinking that often turned to alcoholism. Hawthorne often grumbled about "the uncongenial and sombre weather," "the brooding, murky skies," "the gray gloom," "the spiteful showers," the fog "like slightly rarefied wool." But on the whole Hawthorne felt the English climate had been "shamefully maligned," "its sulkiness and asperities" not nearly so offensive as the English let on. Indeed, Hawthorne liked "the watery sunshine of an English sky," the "gentle rain," the "soft and tender mists." He questioned whether any part of the world looked so beautiful as England on a sunshiny summer day, with "a latent freshness " in the air even then. On such a day, lamented Hawthorne, the English complained of "broiling weather," and hurried to the seaside "with red perspiring faces, in a state of combustion and deliquescence." He found the clear days of September and October "the best weather in the world": a mellow sunshine, a soft, balmy air with a slight haze through it, "exquisite, so gently bright without any glare,—a veiled glow."

Even Taine admired "the eternally fresh grass," the "velvety lawns."

To Hawthorne, England's "sweet and tender foliage," its "exuberant, green luxuriance," its "immortal verdure" were boundless delights. Nothing, commented Hawthorne, was ever allowed to go naked in England; tufts of grass, ferns, ivy, lichens, and "a deep, soft, verdant moss" covered everything. Nature had designed "to mingle a charm of divine gracefulness with even so earthly an institution as a boundary fence."

Both Taine and Hawthorne found "John Bull" a pretty fair representation of the average Englishman. Accustomed to American "paleness and leanness of habit," to America's slender-formed and "haggard" women, to its sinewy and angular men, Hawthorne was a bit offended by "the ponderosity" of the English. He took note of the ample and mottled face of the English male, his "additional chin with a promise of more," and his stomach, which assumed "the dignified prominence which belongs to that metropolis of his system." English girls, with deep, healthy bloom, seemed fitter for milkmaids than for ladies, and by middle age, of enormous bulk, massive with "solid beef and streaky tallow," one thought of them as made up of "steaks and sirloins." The fresh bud of maidenhood grew into an "outrageously developed peony," an "over-blown cabbage rose." However, after several years in England, Hawthorne was driven to the "half acknowledgement" that English women were perhaps "a little finer animals" than American women, who were pale of complexion, thin of voice, "meagre," "scanty," even "scrawny."

Hawthorne admired the English for the great faith they had in their "sturdy stomachs." They ate not recklessly and ravenously, as did the Americans, or exquisitely, as did the French, but deliberately, with due absorption in the business. Even after sixty, most of them were still staunch and hearty eaters, and not given to dyspepsia like elderly Americans. Often a single English meal would consist of trout, ham and greens, pigeon pie, cold beef and mutton, a gooseberry pudding, washed down with bitter ale, "mother's milk to an Englishman." Hawthorne was fascinated by the way the English (many of them too poor to buy fresh meat except on occasion tripe or bullock's feet) gathered around butchers' stalls to gaze at the carcasses of mighty beeves, prize oxen, plump sheep and pigs, festooned with bas reliefs of fat on their ribs and shoulders.

Taine spoke of English "torpor" of nerve and concluded that in England "the physical animal, and the primitive man as nature makes him before turning him over to civilisation, is a stronger, rougher species." To Hawthorne, too, the English had an abundance of animal spirits, an "unpolished ruggedness," a steadiness and "obtuseness" of nerve (less high-strung than Americans), a healthiness and wholesomeness combined with an ingrained coarseness and earthiness. English men and women had a propensity, far greater than the American, "to batter one another's persons." It required a vast deal of refinement to spiritualize such a race. In England a woman was either very much of a lady or she was not a lady. "The sturdy Anglo-Saxon nature does not refine itself short of the third generation," and then "under a surface never silken to the touch," there emerges in the gentleman "a refinement of manners too thorough and genuine to be thought of as a separate endowment." Even so, there was still apt to be a lack of exquisiteness of taste, of subtlety and sense of nuance, of inner élan known to the cultivated of other old societies.

Both Taine and Hawthorne sought for the forces which had chastened this race. Taine found part of the explanation in the prevalence of outdoor sports, which Hawthorne, revealing his own lack of interest in this aspect of life, neglected. Both gave some weight to puritanism, Taine much more than Hawthorne. Both gave great importance to the English class system.

The English, argued Taine, needed all of their great strength to bridle their violent and dangerous appetites, "a Caliban far more savage and far uglier than the merry, jovial satyr of France and Italy." Hence puritanism and the slow, conscious development of private and public conscience—the English strait-laced morality and sense of duty. However, continued Taine, even in Victorian England, one had but to see the uninhibited earthiness of an English holiday crowd to get some conception of the "strong, young sap of the tree" (the rustic revels of Elizabethan England, the carnival of the Restoration, the society depicted by Fielding and Smollett) which puritanism, generation after generation, "so lopped and topped and pollarded that although it finally grew straight it also grew rigid."[1]

While England's puritanism profoundly impressed Taine, what struck Hawthorne, coming from puritan America and even more puritan New England, was the relative persistence of nonpuritanism. After observing

1. Throughout this book, when "puritanism" refers to a set of behavioral attitudes and values derived from a number of cultural sources, the Protestant ethic being but one of them, then the word is not capitalized. (If we are to write Puritan America, then why not write Capitalist America, Free-Enterprise America, Democratic America?) But when "Puritanism" refers to a definite historical religious movement which flourished in the late sixteenth and the seventeenth centuries and to its continuing impact on religious life, then it has been capitalized.

Englishmen in the streets, the parks, and the pubs, their propensity for kissing games, their promiscuous-kissing holiday customs, their vast consumption of spirits, Hawthorne concluded that Englishmen, from peer to peasant, were "certainly a franker and simpler people than ourselves." Over and again he noticed that English men and women did things that would make them ridiculous in America, that they were not afraid of enjoying themselves in their own way. "They adhere closer to the original simplicity in which mankind was created than we ourselves do; they love, quarrel, laugh, cry, and turn their actual selves inside out with greater freedom than any class of Americans would think decorous." Hawthorne recounted an occasion when Charles Dickens, no longer young, acted in a play at Liverpool, spent the rest of the night feasting and drinking and making speeches at table, and wound up in the early morning hours jumping leapfrog over the backs of the entire company. Hawthorne believed that much of the old England of Tom Jones and Joseph Andrews, Humphry Clinker and Roderick Random still survived, that this "singular people" had a certain contempt for "any fine-strained purity, any squeamishness, as they consider it, on the part of an ingenuous youth."

Both Taine and Hawthorne found the class system the most significant aspect of English society. Taine admired this system, felt it gave England the stability France lacked. England's political constitution worked well because it was rooted in social reality. The actual rulers of England were the around 120,000 families of the aristocracy, the gentry, and the old mercantile firms—wealthy, long established, and of local residence—who took the lead in public affairs in the counties and municipalities, and as members of Parliament genuinely represented their constituencies because the local people knew them thoroughly and deferred to them, a fact which made the extension of the franchise of less importance in England than in other countries.

Although he recognized its services, Hawthorne was more critical of the aristocracy than Taine. He wondered if its rewards and privileges were not all out of proportion to its latter-day services. He could not reconcile the vast estates and parks of the nobility with the prevalence of pauperism and almshouses. He was offended at the adulation heaped on the aristocracy. (Even on the walls of the Reform Club in London there were portraits only of lords; not a single commoner had been so honored by the "radicals" who composed the club's membership.) He noted how frequently lords lacked the personal ability and force to live up to the claims of their ancestors, how in the ordinary ranks of life they would be commonplace men. He admired the healthy lives led by the aristocracy and the gentry and their accomplished capacity for making themselves physically comfortable, but he was annoyed at their lack of interest in ideas, at the "alloy" with which they mixed their seeming respect for literature and writers. Few English aristocrats, Hawthorne felt, ever developed the delicacy and tact of the true gentleman. When the Marquis of Landsdowne insisted that Hawthorne take precedence over himself, Hawthorne wrote: "By insisting upon it, he showed his condescension much more than if, when he saw my unwillingness to take precedence, he had passed forward, as if the point were not worth asserting or yielding."

In the course of his four-year stay in England, Hawthorne visited almost every market town in the kingdom, and he was impressed by the way the gentry had made their impact on them: the handsome sixteenth-century and seventeenth-century town houses of the squires, the comfortable inns which reflected the tastes of the local notabilities, the lovely churches made possible by their continuing benefactions. But, again, he recoiled from the deference paid the first families and observed how they took the best pews in the churches and in death monopolized the most prominent burial places, their tombs and monuments dwarfing all the others. He found the gentry overpractical and parochial and relished a story told him by Monckton Milnes of a wealthy squire and prominent member of Parliament who confessed to never having heard of Alfred Tennyson.

What struck Hawthorne about the English rural village was its incongruity. Externally, the homes of the peasants, built of gray stone or brick, with thatched roofs out of which sprouted luxurious verdure, with shrubbery clustering about them, were attractive; but inside, cramped and primitive, they resembled pigsties. Accustomed to the separated and often isolated homesteads of American farmers, Hawthorne found it difficult to get used to the immemorial communal aspects of the English village—the cottages stuck one against another, their thatched roofs forming a single contiguity, all growing together "like the cells of honeycomb," where privacy and self-respect seemed impossible. The peasants were passive; there was no American restlessness or smartness in their minds. The village was insulated; there was no news of the outside world; the villagers had no interest in public affairs; they regarded

197

such matters as the exclusive concern of their betters. (Hawthorne found the rural villages of Scotland still drearier; even the old churches had been "uglified" and "kirkified" with whitewash.)

Taine and Hawthorne were in agreement in their estimate of rural poverty. Taine held that English peasants were more wretched and backward than those of Continental Europe. What added to their plight, said Taine, was the slovenliness of the English peasant women, their wanting in address, their utter lack of any sense of household management, of making-do. Every member of a peasant family could hoe a field; not one could mend a coat or cook a turnip.

But among the rural residents of England above the peasants, those outside the squalid villages, those living in individual cottages—yeoman farmers, rectors (who drew the livings), curates (who did the work), schoolmasters, pensioners, shopkeepers—life had many aspects of pastoral delight. This was clean, fresh England; sweet, peaceful England; the England of polished copper cups and burnished teakettles, of flower-bordered cottages, ivy walls, moss-covered flagstones, shaded lanes, and trimmed hedges. This was Wordsworthian England, most of it taking its cues from the squires in the big houses. Hawthorne had much more "feel" for this aspect of England than Taine.

Contrasting England's wealthy, urban mercantile and industrial entrepreneurs with those in America, Hawthorne found that the former were paternal in their relations with their employees, less speculative in business, and more given to civic causes, public subscriptions, and grand ceremonial dinners. As for the less wealthy, the middling business classes, Hawthorne described these as decorous, earnest, eager for instruction and self-improvement, "ponderously respectable," close and exacting in money matters, more timid and less venturesome than their American counterparts, but better housed and with a greater knack (like all other Englishmen of means) for solid comfort.

Two groups of Englishmen surprised Hawthorne, for they seemed scarcely to exist in America. One was the *rentier* class living on low incomes, the retired and the pensioners, comprising "a second-class gentility," found in abundance at the numerous spas and coastal resorts. The other was the large number of old people of all classes, who gathered in the streets and around the churchyards. There was an acceptance of age, no shame in growing old, no artificial contrivances of "a skin-deep youthfulness," as in America.

Not even Dickens has left more ghastly descriptions of the city poor, the other half of the two Englands, than Hawthorne. There was the upper poor, with enough means to wash and have a Sunday change of clothes. There was the lower poor, who seemed never to wash and wore the same tattered clothing on work days and holidays. And there were the beggars, numerous enough to form a separate class, who resisted the workhouses for the city streets.

The lower poor, deserting their cramped, filthy, stinking quarters, lived largely in the chill of the streets, often went barefooted in the slosh of winter to save back their shoes, scrambled for the coal and vegetables that fell from overladen carts, and their ragged children and sluttish women, with old shaving mugs and broken-nosed teapots, lined up at the gilded door-posts of the gin shops, "the spirit-vaults," to carry home their potions of "madness." The poor seemed a race apart; their children, even when sturdy, had "mean, coarse, vulgar features and figures betraying a low origin and ignorant and brutal parents." And the gloom, the dampness, the fog, and the coal smoke cast over the districts of the poor an "adhesive grime," a "mourning garb," somberly "phantasmagoric," "fascinating in its ugliness." Hawthorne commented that even the poor-Irish sections of Boston were nothing like this. He would have agreed with Taine that "the vile and the horrible are worse in England than elsewhere."

Most impressive was the way the English poor, an earthy people, who in holiday mood thought nothing of pinching the buttocks of strangers in their streets (Taine) and who would pummel the daylights out of one another at the drop of a hat (Hawthorne), accepted their lot with patience and meekness, and gaped with respect and awe at a fashionable wedding party or the equipage of a lord.

With his eternal fascination for the bustle of city streets, Hawthorne recorded vivid descriptions of beggars and knaves, itinerants and performing vagabonds, fiddlers and buglers, ballad singers and folk-dancers, Highlanders with bagpipes and lassies dancing hornpipes, accordionists and organ-grinders, Punch-and-Judy and town criers, hawkers of hot snails and hags selling from the pavements half-decayed oranges and cheap trinkets. And when the fog settled over the Mersey or the Thames and scattered along the vistas of the streets, the misty glows, from the steamers, the coal furnaces, the gas lights, the lanterns swinging from pony carts and drays, and the charcoal fires of women and boys roasting chestnuts at street corners, etherealized the material, making the scenes resemble "the other world of worldly people, gross even in ghostliness."

Cultural Ethos and Technocracy

Gradually Hawthorne became aware of certain advantages of the class system. Among the poor, he often discovered a charm and simplicity, a naturalness never vulgarized by attempts to assume manners and adornments of another station. There was no striving, as in America, to become "genteel" and "lady-like." Again, the poor and even the beggars had status, their rights were respected. In his comings and goings across the Mersey, Hawthorne noticed many times how the poor and even the beggars took the best places on the ferry's deck, appropriated the seats closest to the stove in the cabin, and without self-consciousness ate their snacks in the presence of the other passengers. They were never rudely treated, never made fun of, never looked down upon. Servants, too, were a class apart; they had well-defined status and prerogatives. Hawthorne was told that Thackeray's sensitivity in the presence of servants amounted almost to a dread of them. Servants expected their betters to observe the amenities, and Hawthorne learned that when Carlyle appeared at social gatherings in shabby dress, as he sometimes did, flunkies could not resist responding to this species of rudeness with a kind of rudeness of their own.

On the whole, different classes seemed to encounter one another in an easier manner than in America; the shock was less palpable, for the differences were real and there was no need to assert them. Despite the class amenities adhered to, there were fewer pressures to conformity than in America and more tolerance of odd behavior and individual eccentricity. There was not the drive to homogeneity that there was in America; being used to class differences helped accustom people to individual differences; both the upper classes and the lower classes were less motivated by middle-class urges to condition individual conduct in order "to get ahead"; the upper classes had obviously "arrived," and the lower classes were not likely "to arrive" no matter what they did. The class system stabilized British society, allowed institutional changes to be made without violence and by slow degrees and to be assimilated into the warp and woof of the traditional past.

However, the American in Hawthorne would not allow him to ignore the obvious disadvantages of the class system. He discovered that even those who broke through the class barriers and achieved a self-made success were not genuinely at ease, whereas in America a self-made man came naturally to distinctions and indeed took them for granted. Hawthorne recounted how he was taken to dinner in the House of Commons restaurant by a member who had risen in his profession and in politics despite his low birth. But his host was no more a gentleman than when he began life, not a whit more refined inwardly or outwardly; he took a childish delight in his position, continued to wonder at it, could not believe it was real after all.

Hawthorne himself was a victim of the social formalities of nineteenth-century England, for despite his literary fame, his official position, and his many friends, he was prevented by the amenities from meeting the English literary men he most wanted to know. It was relatively easy for a visiting Englishman, celebrity or no celebrity, to meet a Hawthorne, an Emerson, a Lowell, or a Longfellow in America; but Hawthorne spent four years in England without ever meeting Thackeray, Dickens, Carlyle, or Tennyson. The opportunities did not arise naturally, and it would have been bad form to have sought them deliberately. On one occasion, with ill-concealed excitement, Hawthorne trailed Tennyson through the rooms of a public art exhibition, taking note of the poet's every expression and gesture, but he dared not defy the conventions and introduce himself.

However, Hawthorne did come to know, in one way or another, Leigh Hunt, Harriet Martineau, Barry Cornwall, Douglas Jerrold, Coventry Patmore, and others. He was drawn to "the fineness, subtilty, and grace" of Hunt, but Miss Martineau comes off as a kind of Dorothy Thompson with an ear trumpet. Through Monckton Milnes, whose famous breakfasts and dinners brought a larger number of interesting people together than any other hospitality in nineteenth-century England—in the following decade they would bravely attempt a pairing of Henry Adams and Algernon Swinburne—Hawthorne once met Macaulay and the Brownings; and through Milnes and other hosts he often sat at table with many of the transient celebrities of the day, including that formidable wit Tom Taylor.

English antiquity, too, along with the class system, explained the chief differences between the English and the American society. Unlike Taine, who came from a country older than England, Hawthorne was struck by the omnipresent evidences and musty odors of age, the venerability of things and institutions, the sanctity of even a turnip field. The humanist in Hawthorne was attracted but the restless American in him repelled by the seeming immobility—the wearisome round of living from century to century in the same place, going in and out of the same doors, cultivating the same plots, meeting the same faces, marrying one another over and over again, going to the same church, lying down in the same graveyard.

Cultural Ethos and Technocracy

Although Hawthorne "scarce ever found an Englishman who seemed really to desire change," the miracle of England was the way incongruities were harmoniously patched together, modern fronts made to hide antique arrangements, new beams fitted into old ones, not only in physical things, but (and here Taine was impressed) institutionally. Taine commented: "In England the new generation does not break with its predecessor; reforms are superimposed on institutions; and the present, resting on the past, continues it."

Even the shock of the Industrial Revolution, experienced first in England, without any past guideposts from English or any other society, was being successfully cushioned by English conservatism; even the fearsomely dynamic forces sweeping out of "Pluto's sulphurous cities" were being chastened by English traditionalism. This indeed was a social maturity which both Taine and Hawthorne were compelled to respect, although Hawthorne was somewhat less sure than Taine that England would entirely escape revolution.

Both a cause and an effect of this capacity for reconciling the new and the old was the remarkable consensus among both Conservatives and Liberals about basic values. Hawthorne wrote that "a miscellaneous party of Englishmen can always find more comfortable ground to meet upon than as many Americans, their differences of opinion being incomparably less radical than ours, and it being the sincerest wish of all their hearts, whether they call themselves Liberals or what not, that nothing in the world shall ever be greatly altered from what it has been or is." But Hawthorne was writing before the Reform Bills of 1867 and 1884, in the period when both the English parties still represented a small ruling class and were led by the aristocracy, and he doubtless was thinking, too, of the asperity of the party contests in the United States during the 1830s, 1840s, and 1850s, when the fierce substantive battles of Jacksonians and anti-Jacksonians were closely followed by the bitterness of the slavery debates.

However, Hawthorne felt that there was something beyond this, an American virulence of party conflict per se, which was happily absent in England. He lamented that Americans regarded government as merely a party phenomenon, that they cared nothing for the credit of it unless it was the administration of their own party. In England, on the other hand, all people, of whatever party, seemed anxious for the credit of their rulers. Too often Americans thought of their government as a mere collection of partisans, self-seeking and parvenu, that changed hands every four years, and hence the government rarely commanded general American respect.

The century intervening since Hawthorne wrote has produced many changes. One of the most obvious is the decline of heavy eating in England and the slenderizing of English forms. However, it is noteworthy how many of the distinguishing folkways and little habits of national life, mentioned by Hawthorne a century ago, still persist. The Americans still overheat their houses in winter, the English still underheat theirs. (Hawthorne came to feel that the Americans *did* overheat their houses.) The Americans still overlight their cities, the English still underlight theirs. (Hawthorne came to prefer the more subdued English cities to the "glare" of American cities.) The English are still less inclined to formal organization, while the Americans, even more than in Hawthorne's day, display "that native tendency to organization." (Hawthorne remarked that even when several Americans paid a visit to the consul, they would halt at his door, hold a meeting, and elect a chairman of the delegation.) The "up-and-down intonation" of English speech still strikes the American as affected, while the English still feel that American speech is flat. (Even in Hawthorne's day Americans curiously referred to Englishmen "as speaking English with an English accent," and Hawthorne had his own pet device for determining whether a person was English or American: if he pronounced "been" as "green" then he was no American.) The English still think of American music as Negro music; jazz and its variations are popular in England, just as in Hawthorne's time Southern Negro melodies were about the only American music the English knew. Americans still avidly read of the goings-on of English royalty, a stock in trade today of American women's magazines of mass circulation. (However, Americans no longer dream of inheriting English titles, as they did in Hawthorne's time, when he spoke of "this peculiar Anglo-American insanity," although it persisted into the following generation sufficiently for Americans to take Little Lord Fauntleroy to their hearts.) English journalists, like the journalists of other countries, are still notorious "free-loaders," and today, as in Hawthorne's day, every Englishman runs to *The Times* with his little grievance.

The English continue to make political and social changes gradually and peacefully, to harmonize the new and the old with a minimum of conflict. Since Hawthorne's time the English have developed universal suffrage, the welfare state, and a semi-socialist economy, and they have made impressive inroads on poverty and class privileges—all without violence, the only major country of Europe to escape the revolutions of the nineteenth and twentieth centuries. (Although Hawthorne emphasized the English consensus on basic social and political values, there was a time, during the Crimean War, as popular discontent rose high, when he felt England might face revolution and the overthrow of the aristocracy and even the monarchy. But instead that war helped hasten, during the following two decades, the peaceful extension of the suffrage and the enactment of the social reforms known as "Tory Socialism." Incidentally, Hawthorne's journal reveals how little the ordinary lives of Englishmen were actually affected by the Crimean War, but even this nontotal nineteenth-century war stimulated drives to institutional change. It is a wonder that conservatives anywhere *ever* favor war, for war has been the great change-maker of history.)

Today, it is the Americans who are considered to have the wider consensus about basic social and political values, while the English, divided between Conservatives and Laborites, capitalists and socialists, display broader and deeper political differences than the Americans. However, even under these altered circumstances, the English continue to carry on their party contests in a spirit of restraint and moderation.

One of the most marked changes since Hawthorne's day is that the Anglo-American ambivalence now emphasizes its friendly rather than its hostile sides. When read today, the bristling evidence, revealed by the journals, of Anglo-American antagonism a century ago must come as a shock to both Americans and Britons. Hawthorne noted that a "cold, thin medium" intervened between even the most intimate personal relations of an Englishman and an American. As for the generality, Hawthorne confided: "If an Englishman were individually acquainted with all our twenty-five millions of Americans, and liked every one of them individually and believed that each of these millions was a Christian, honest, upright, and kind, he would doubt, despise, and hate them in the aggregate, however he might love and honor them as individuals."

The English regarded the Americans as uncouth; conspicuous rudeness by anyone was labeled "Yankee conduct"; America was characterized as an unhealthy land of sickly people. Englishmen took a peculiar delight in belittling American heroes and puncturing American "myths"—for instance, among the many "digs" to which he was subjected, Hawthorne was told that the Mayflower, after carrying the Pilgrims to Plymouth, was thereafter employed to carry slaves to the West Indies ("nuts for the Southerners," commented Hawthorne in his diary), that John Paul Jones, while in the English service, was such a poltroon he had to be flogged publicly.

At times Hawthorne reacted with some asperity of his own. "These people think so loftily of themselves and so contemptuously of everybody else, that it requires more generosity than I possess to keep always in a good humor with them." During the Crimean War, he wrote: "I seem to myself like a spy or a traitor, when I meet their eyes, and am conscious that I neither hope nor fear in sympathy with them, although they look to me in full confidence of sympathy." Again: "I shall never love England until she sues to us for help, and, in the meantime, the fewer triumphs she obtains the better for all parties. An Englishman in adversity is a very respectable fellow; he does not lose his dignity, but merely comes to a proper conception of himself."

Conscious of English prejudices about Americans, Hawthorne leaned backward not to confirm them. Sometimes he was over protective. He blushed to think that Minister James Buchanan, in the presence of Englishmen, would take out his big colored silk handkerchief and tie knots in it. He was even more uneasy about the impression Buchanan's successor, George M. Dallas, "a humbug," was making on the English. (Not that Hawthorne was devoid of that "noble weakness," patriotism; his heart swelled when political exiles from Poland, Hungary, and Napoleon III's France beat a path to his consulate to beg passage to "the blessed land of Freedom.")

The truth is that Hawthorne loved England very much. Many of his strictures on the English were defensive. From the beginning Hawthorne felt at home in England, the longer he stayed the better he liked it, and he wrote that even if America went down he would not be a complete stranger in the world as long as there was an England. "I never came into personal relations with an Englishman without beginning to like him, and feeling my favorable impression wax stronger with the progress of the acquaintance. I never stood in an English crowd without being conscious of hereditary affinities." Again: "What a wonderful land! It is our forefathers' land; our land, for I shall not give up such a precious inheritance." Hawthorne felt that every society had a consistency and harmony when viewed from the inside; he deplored the fact that Englishmen who wrote about America seldom stayed long enough to discern the inner truth about the American society; he hoped that his own observations about the English would not be marred by similar failings. But when his *Note-Books* was published, the English felt they had been grievously misjudged.

Hawthorne made little mention of the reasons for the Anglo-American antagonism of his time, but these are not difficult to discern. The Americans were still close to the Revolution, the English-inspired Indian wars, and the War of 1812, and on many fronts they were still made to feel the barbs of English hostility and arrogance. On the English side, the very existence of the United States was a standing affront. The colonies had won their independence not in an age of British decline but at a time the British were on the rise as an imperial and a world power. Save for the American Revolution, the eighteenth and nineteenth centuries witnessed a steady expansion of British greatness. America was the one major reverse, a reminder that Britain was not infallible. It was all the more galling that the one noteworthy British backset should have been administered by their own kinsmen, in a revolution which really need never to have happened. As America grew in wealth and power, the disaster seemed all the greater. In characteristic human fashion, the English sought to relieve their inner sense of chagrin and remorse by belittling the Americans.

The Hawthorne revealed by the diaries was in many ways different from the old stereotype of him. He was modest, but he had the literary man's usual occupational vanity. He noted the presence and precise edition of any of his books in a shop, library, or home; he was obviously pleased when Englishmen spoke to him of his works; he was amused when an Englishman asked if he were not the author of "The Red Letter A."

Hawthorne was shy, but not nearly so much as was supposed. During his years in England, this "fanciful recluse" (the term he himself used to describe the reputation he bore among his countrymen) became surprisingly gregarious. English conditions, much more than American, were suited for bringing Hawthorne into association with others. He liked the breakfasts and dinners where men of affairs and ideas conversed for hours at table. In the beginning, he was frightened almost to speechlessness when called on to respond to a toast, but with experience he came positively to enjoy the declamatory sound of his own voice, and he reached such a degree of effectiveness in public address that eventually it became usual for him to get more "hear-hears" and applause than any of the other speakers. On one occasion he sat on the edge of his chair, eagerly wanting to make the speech on the house of Stanley which the scion of that house, later Lord Derby, was so lamely making. However, Hawthorne thought all oratory was humbug, particularly the spread-eagle, stemwinding variety then in vogue in the United States.

Hawthorne's mind was fundamentally religious. Sin and suffering had a horrified wonderment for him, as they had for St. Augustine and his own Puritan forebears. How could such things be in a God-made world? He confessed that he looked at human misery and depravity with the same fascinated revulsion, the same "loathsome interest," although to a far intenser degree, that he felt when, as a boy, he turned over an old log that had lain on the damp ground and found "a vicious multitude of unclean and devilish-looking insects scampering to and fro beneath it." Walking through the streets of the poor, he wondered how anything so precious as a germ of immortal growth could be buried under such a dirt heap, such a cesspool of misery and vice. And then doubt; for if the children of the wretchedly poor were to have no immortality, what superior claim could he have for his?

Hawthorne was amazed at the growth of spiritualism, and annoyed that mediums and ghosts should so frequently be the topics of dinner conversation. Even much of his table talk with Elizabeth Browning was about her interest in spiritualism, and he was puzzled that so delicate a mind should be attracted to such rubbish. Hawthorne believed that spiritualism represented a decline in the religious imagination, a long step to materialism and secularization. On the other hand, Hawthorne was no mystic; unlike his friend Herman Melville, he was not given to speculation about "Providence and futurity, and everything else that lies beyond human ken."

While the fundamental impulse in Hawthorne was religious, even Puritan, his "awful power of insight" (to use a friend's characterization) was mainly concerned with the psychic, the social, the human condition; it was not abstract, even less theological. What he probed was religion as an expression of man's inner self and the effects of religion, in turn, on that inner self. This was what gave Gothic churches—made out of "the dim, awful, mysterious, grotesque, intricate nature of man"—an endless fascination for him.

What Hawthorne liked about Puritanism was its emphasis on ethical content and the centrality of the sermon in the religious service. What he disliked about Puritanism was its rejection of ritual, its frequent insensitivity to beauty and church architecture, its drastic oversimplification of human nature, and its lack of compassion. Hawthorne could never find it in his heart utterly to condemn England's gin-drinking poor for seeking to lift themselves out of the smothering squalor of their lives, and he felt that the temperance reformers had not been taken fully into the counsels of the Divine Beneficence. As for the other vices, Hawthorne believed that the more open avowal of immoralities in England than in America "served to throw the disease to the surface, where it might be more effectively dealt with, and leave a sacred interior not utterly profaned, instead of turning its poison back among the inner vitalities of the character, at the imminent risk of corrupting them all."

Hawthorne was not a melancholy or morbid man. Quite the reverse. His mind was healthy, his disposition cheerful, his temperament equable. He derived aesthetic and intellectual enjoyment from exercising his insights into the incongruities and ambiguities of the human condition. He wrote over and again of his "sombre satisfaction" and his "sombre pleasure" in probing into the dark corners of society and of the human mind. Mrs. Hawthorne has written (and I cannot escape the impression she was quoting from a suppressed passage of the journal in which Hawthorne described himself) that her husband "saw too far to be despondent, though his vivid sympathies and shaping imagination often made him sad in behalf of others. He also perceived morbidness, wherever it existed, instantly, as if by the illumination of his own steady cheer; and he had the plastic power of putting himself in each person's situation, and of looking from every point of view, which made his charity most comprehensive." In his journal, under date of Christmas 1854, Hawthorne wrote categorically that he considered himself famous, prosperous—and happy, too.

Hawthorne had a warm heart, and a cold heart repelled him. Granted his many references to warm and cold hearts reflected the romanticism of his time, still Hawthorne's comments have a ring of deep personal feeling. Among those whom Hawthorne regarded as having cold hearts were Lady Byron and Harriet Lane, Buchanan's niece and his hostess in the London legation and in the White House. After hearing that Thackeray read without emotion the saddest portions of *The Newcomes* to friends in a cider-cellar, Hawthorne wondered if Thackeray would not have to be consigned to the category of the cold-hearted. "Speaking of Thackeray, I cannot but wonder at his coolness in respect to his own pathos, and compare it with my emotions, when I read the last scenes of *The Scarlet Letter* to my wife, just after writing it,—tried to read it rather, for my voice swelled and heaved, as if I were tossed up and down on an ocean as it sub-sides after a storm."

When on the eve of the publication of *Our Old Home*, in the midst of the Civil War, Hawthorne was beseeched to withdraw his dedication to Franklin Pierce, then fanatically unpopular in the North, he replied that if he were to tear out the dedication he could never look at the book again without remorse and shame. Thus in the very act of publishing, Hawthorne revealed other personal characteristics—loyalty to friends and concern not for his public image but for the integrity of his own soul.

❧ The Passing of Bughouse Square

The colorful folk gregariousness and merriment of Europe's city streets, so vividly described and so much enjoyed by Hawthorne, had their nearest counterparts in America's bughouse squares, which reached their height in the nineteenth and early twentieth centuries, when they operated as a social outlet and even catharsis for European immigrants, hungry for the *joie de vivre* of their homelands. These squares were always much more than that, but they would have been impossible without the newcomers from Europe.

All through our history, European visitors have been impressed by the lack of uninhibited public gaiety in America. Charles Dickens was shocked at the quietness of the New York streets. Where were the itinerant bands, the folk dancers, the wind and stringed instruments, the Punches, fantoccini, dancing-dogs, jugglers, conjurers, Orchestrinas, and barrel-organs? Not so much as a white mouse in a twirling cage, he lamented. Alexis De Tocqueville found the Americans a people of "astonishing gravity." He observed: "An American, instead of going in a leisure hour to dance merrily at some place of public resort, as the fellows of his class continue to do throughout the greater part of Europe, shuts himself up at home to drink. He thus enjoys two pleasures; he can go on thinking of his business and can get drunk decently by his own fireside. I thought the English constituted the most serious people on the face of the earth, but I have since seen the Americans and have changed my mind." Would the American have found a lively gregariousness even had he deserted his solo drinking for the tavern? No, for Europeans found American tippling places downright sullen. Anthony Trollope commented that patrons of American taphouses often avoided conversation, sat chewing their tobacco cuds in silence, and just ruminated.

Bughouse squares in America, then, were unique in their public folk gregariousness; when once going concerns, they survived the virtual cutoff of European immigration; and it was not until the 1950s, when the impact of the technocratic, affluent, and suburbanite civilization became noticeable to all, that bughouse squares, little and big, suffered a precipitate decline.

Even had bughouse squares otherwise survived in full vigor, it would have been difficult to keep alive the old animated ideological

Antioch Review, XX (Fall 1960).

debates. Differences between Democrats and Republicans had been further blurred by a nonpartisan foreign policy. Although rightism came to the fore during McCarthyism and then again during the Goldwater campaign, the many and various leftist groups in America had so declined by that time that the debates would have lost their old pungency and tang. Merely resisting rightism does not carry the excitement of opposing it with a variety of positive leftisms.

Since the following article was written, the new youth movements and cults have arisen. These are asking some questions about the meaning of human existence, and they are seeking a new nonconformity and personal autonomy. But often these movements unconsciously exhibit the very technocratic values against which the young rebels fancy themselves in revolt. A New Leftism has emerged, too, but this is mostly confined to the young, is highly romantic, and does not invite disputation. All too often the New Leftists spurn discussion and substitute epithets and obscenities for debate. The new youth movements are discussed in the concluding article of this book.

Are hippies, leather-jackets, and New Leftists developing new bughouse squares? Here and there, as at Dupont Circle in Washington, there is some indication that they are. But the weight of the evidence is against this. The old bughouse squares were broadly inclusive; people of all ages and conditions gathered there; an atmosphere of folk carnival prevailed; the ideologies, cults, and fringe movements represented were numerous and wide-ranging; vivacious discussion and debate were invited and even demanded. On the other hand, today's rebellious youth gather less at plazas and more at small bistros and hangouts in "emancipated" neighborhoods; their activities are mostly confined to young people, and oldsters and outsiders are not welcome; the cults represented are few and highly stereotyped; the point of view of each cult is taken for granted; there is not much tolerance of genuine discussion and debate. The difference is that of the Pershing Square of yesterday and the Hollywood "Strip" of today.

Incidentally, Pershing Square, in these days of its decadence, has been officially rebaptized MacArthur Square.

■ AMERICA'S COLORFUL and zany bughouse squares are on the way out. Before long they will have disappeared as completely as vaudeville, personal journalism, the *bon vivant* of the boulevard, the Bohemian, and the old-fashioned saloon with its sawdust-strewn floor and free lunch. A recent leisurely trip over the country has brought home to me this unhappy fact.

Bughouse squares have not entirely vanished from the American scene,

Cultural Ethos and Technocracy

but they are slipping badly. Some have disappeared without a trace, and in those which have survived there are fewer participants, the debates and discussions are less vital and diverse, and in some of them a grotesque exhibitionism proclaims their decadence.

Now, what is a bughouse square? A bughouse square is a public street-intersection or mall or park in a large city where gregarious, imaginative, exhibitionistic, and auto-compulsive "ism" peddlers, agitators, soapbox-ers, folk evangelists, teachers, showmen, faddists, cultists, cranks, crack-pots, dreamers, and self-proclaimed messiahs congregate to impress one another and to display their wares. These performers attract a large number of disciples, camp-followers, hecklers, "wise-guys," honest inquirers, and humble seekers. There is always present a large number of the skeptical and the disputatious. And there is invariably a sprinkling of unemployed, hobos, punks, pimps, faggots, fey proletariat, strumpets, beatniks, and the more robust and less fastidious of the Bohemians. In short, a really first-class bughouse square is a bit of skid row, carnival, evangelistic revival, poor man's town hall, and people's university rolled into one.

Bughouse squares have existed since the dawn of urban civilization, from the time some Socrates held forth in a city-state market place to London's amazing Hyde Park of today. Readers of Macaulay will recall his vivid description of a lusty bughouse square of around 1700, London's Lincoln's Inn Fields, where every evening the poor congregated, agitators harangued, and trained bears danced. The prerequisites for a first-class bughouse square have been a concentration of population, metropolitan-ism, a meeting and mixing of diverse experiences and cultures, and a degree of freedom—cultural freedom even more than political freedom.

Some of America's early city squares became bughouse squares. Independence Square was long a center of popular agitation. Boston's famous Common was at its height in the 1840s when the oddities holding forth there were abolitionists, Fourierites, women's rights advocates, and Amelia Bloomer "in her most extreme costume—not skirts at all but trousers, full and stiff, reaching to her ankles and tied there." New York has had a long and varied history of bughouse squares: first Bowling Green, then City Hall Park in the 1830s and 1840s, then Tompkins Square in the 1870s, when laborites, antimonopolists, and greenbackers held sway. By the turn of the century Union Square was coming into its own.

The most famous of America's bughouse squares during the twentieth century have been Union Square and Columbus Circle in New York, Washington Square in Chicago, and Pershing Square in Los Angeles.

Union Square was in its heyday when I first visited it on the night of August 14, 1916. That visit was the climax of a memorable day for me, a wide-eyed youngster from Indiana on the first day of his first visit to New York City. At that time the Square was not so exclusively political and left-wing as it was to become in the 1920s and 1930s. To me, fresh from a provincial city in Indiana, the spectacle of the Square was outrageously exciting. The nearest thing to it in my experience had been courthouse square on a Saturday afternoon in October of election year, when Indiana farmers gathered and heatedly debated the merits of Republicans, Democrats, and Bull Moosers. But here in Union Square were many soapboxers, curbstone agitators, and "spontaneous" disputants, around whom were gathered knots of hecklers and listeners, holding forth on the most strange and forbidden subjects: pacifism, feminism, birth-control, vegetarianism, the single tax, agnosticism, atheism. The left-wingers were bitterly divided between Big Bill Haywood's Wobblies, Daniel De Leon's Socialist-Labor party, and the Socialists of Debs, Benson, Hillquit, London, and Berger, who were themselves divided.

Since that night, I have always made it a point to visit Union Square on my many trips to New York. I witnessed, in 1920, the protests over New York's Lusk "gag" laws and over the expulsion of the Socialist assemblymen from the New York Legislature. I saw at first hand the decline of the Wobblies and the Socialists and the rise of the American Communists, and how more and more the Communists took over the Square. From the debates and activities in the Square I was able to follow more intelligently the zigzags in the party line, the rise and fall of factions in the Kremlin and how these reflected themselves in the rise and fall of Communist leaders in America. I was on the Square on that astonishing day when the Hitler-Stalin Pact was announced and saw the Communist big-shots, for once at a loss to cope with the situation, slink away from the Square while many of the disillusioned openly wept bitter tears.

Since 1916, the history of the Square may be summarized something like this: gradually it became more political, then more left-wing in general, then more Socialist, then more Communist. With the decline of the American Communists, much of the Square's vitality has departed.

Meantime, a more inclusive bughouse square was developing farther uptown at Columbus Circle. By the 1930s Columbus Circle had attracted most of the nonleftist activities once centered in Union Square. In this period Columbus Circle was much more like London's Hyde Park than was Union Square. Religion, philosophy, science, literature, and art became leading subjects in Columbus Circle, although no American bughouse square has had the amount of literary and artistic discussion common to Hyde Park. (Skeptics might insist that pseudo-religion, pseudo-philosophy, pseudo-science would be more accurate descriptions of the subjects discussed, but this would not be quite fair.)

At the Circle, at four o'clock on a Sunday afternoon or at seven o'clock in the evening on a weekday, up would go the sawed-off ladders which served as soap boxes, and on the top of the last rung would be hoisted a small flag, cross, or other symbol of the speaker's subject. By eight or nine o'clock as many as ten or twelve ladders would have speakers, around which clustered the faithful, the skeptical, and the curious. The speakers would be variously engaged in expounding Catholicism, Judaism, Methodism, fundamentalism, modernism, agnosticism, atheism, antisemitism, anti-"popery," free will, determinism, relativity, and so forth. Because the leftists monopolized Union Square, the right wing came in increasing numbers to Columbus Circle. By the late 1930s isolationists, America Firsters, Coughlinites, Silver Shirts, Christian Mobilizers, Bundists, outright Hitlerites, and sundry denouncers of Jews and "Jewish international bankers" came to occupy a large number of the ladders.

During the years immediately preceding Pearl Harbor, the crypto-fascist allies more and more took over Columbus Circle. Antisemitism became rampant. However, those they assailed, including the Jewish people, were not intimidated; they continued to appear at the Circle. Some of the ladders had speakers who denounced the fascist fringe. The mood became ugly and menacing.

A notoriously vulgar agitator was an expressionless and hard-bitten shrew called Fatima, from the pantaloons she habitually wore. She would march to her ladder surrounded by a moronish-looking claque bearing Christian crosses and American flags. Obscenities against the Jews poured from her as smoothly as water out of a jug, but she appeared completely insensitive and unfeeling, unconscious of the vile rottenness she spewed. Her performance was trancelike, and as so often happens in such instances, it cast a kind of hypnotic spell over her listeners. She usually drew the biggest crowd of the evening, and she attracted sufficient attention to make the news columns of some of New York's leading daily newspapers.

One evening I stood right next to Fatima's ladder in such a way that I could also face her audience. At an opportune moment I broke in on her harangue with one of my own. Appearing to address my remarks to Fatima, my harangue went something like this: "Were you born in New York City? Do you know anything about the rest of the United States? Are you not a New York yokel? Did you ever hear of the Ku Klux Klan? Don't you know that in most of the United States the people who hate the Jews also hate your claque here, the Coughlinites, because they are Catholics? Are you trying to stir up the ugly spirit of the Klan again? When you arouse native American prejudices against the Jews you also arouse them against other minorities, particularly the Catholics. Are you so ignorant of America that you do not know these simple facts? Yet you prate about your Americanism!"

The reaction to my own little speech was mixed. At first the crowd was hostile, then incredulous, then silent, and then on the fringes came hearty applause and cries of encouragement. Fatima froze to her ladder. She became tongue-tied. Then she began to shout shrilly to me to get my own ladder and my own crowd and not steal hers. But the jig was up for Fatima that night. Her claque broke into singing, raised aloft their crosses and flags, helped Fatima from her ladder, and ingloriously stole away. Meantime, I had been surrounded by several dozen supporters, many of them sturdy Jewish youths. They constituted themselves a kind of bodyguard. They expressed animated gratitude, but implied I must indeed be a provincial, an innocent, not to know the danger I ran from Bundist thugs. Several of us adjourned to a nearby bar and for hours gloomily discussed the sad state of affairs in the world of that summer of 1940.

To Chicago goes the distinction of giving these folk squares their generic name, for Chicago's Washington Square has long been known by Chicagoans as Bughouse Square. Chicago's Bughouse Square is dark and dismal, forever shadowed by the brooding bulk of the Newberry Library. It has had a greater touch of skid row about it; there you would find more unemployed and more tramps and hobos than in the other bughouse squares of the country. The Wobblies were stronger in Bughouse Square than in the other squares, and even today these ghosts of the past are in evidence there.

Chicago has not escaped a certain provincialism with respect to her

Bughouse Square. At the height of the summer "season," from about nine to eleven in the evening, "rubberneck" buses would stop before the Newberry Library—sometimes as many as one every ten or fifteen minutes—to let visiting Rotarians, clubwomen, and college sorority girls from Kokomo, Kalamazoo, and Keokuk gape at the soapbox revolutionaries. The expressions on the faces of the visitors—ranging from amusement, to incredulity, to fear, to anger, to disdain, to disgust—were no less a spectacle for the Square's habitués. The soapboxers took a malicious glee in further horrifying the gapers and would cast outrageous obscenities in the direction of the buses.

By all odds the giddiest bughouse square in the country has been Pershing Square in Los Angeles. There are those unkind enough to intimate that the Square has been but a microcosm of the city itself. Pershing Square has been not only the most colorful and screwball of the folk squares, but it has also been the most unspecialized and most diverse. The bughouse activities which in New York were divided among Union Square (the leftist), Columbus Circle (the rightist and the religious), and Greenwich Village's Washington Square (the Bohemian) were in Los Angeles all concentrated at Pershing Square.

Pershing Square has represented the uprooted, the restless, and the polyglot population that is Los Angeles. People of vastly diverse origins gathered there: native Californians, transplanted Europeans and Americans, Mexicans, and various kinds of Orientals. The descendants of Kansas and Iowa abolitionists mingled with the Jim Crow boys from Texas and Oklahoma; Southern Negroes became inoculated with the independence and verve of Northern Negroes. The West was always there—cowboys, miners, and migratory laborers.

Los Angeles' famous square has had a larger number of cultists, faddists, fakers, faggots, primitive evangelists, and self-proclaimed messiahs than the other squares. There was more hymn-singing and folk-praying. There was more frivolous heckling. Speakers were interrupted with catcalls and cries of: "Have you had three square meals today?" or "Why don't you hang the hammer and sickle around the Pope's neck?" There were more capers and high jinks. A decade ago there was a Negro youth in turban and red cape who regularly came to the Square to do a ballet, turning off cries of "fruit cake" with banter and smiles. There were players and singers with banjo, guitar, and mandolin. There were cowboys who strummed and sang the plaintive ballads of Arizona mesa and Texas plain. There were those heartwarming community sings in which Negro and Jim Crow buried their smoldering hates and united in lusty renderings of "Dixie," "Old Black Joe," and "John Brown's Body."

Pershing Square was most diverse in its politics and religions. Two decades ago all the well-known varieties of Marxism were there, and in addition any number of splinter offshoots never heard of in Europe or Eastern United States, offshoots which represented a strange mixture of Marxism and the American frontier. All the rightist movements of Columbus Circle were there, but the followers of Gerald L. K. Smith seemed most numerous. All of the traditional religions and sects of America, the majority ones and the minority ones, were represented; but there were also Jehovah's Witnesses, Seventh Day Adventists, Mormons, followers of Father Divine, New Thoughters, votaries of One-World religion, theosophists, Rosicrucians, prophets claiming new revelations and personal miracles, messiahs with the short beard of the Jesus tradition proclaiming their virgin births, foes of "Churchianity," iconoclasts, agnostics, earnest young doctrinaires parroting conventional atheistic arguments, Communists disguised as Catholics better to proselytize for Communism, trollops preaching religion and hoping for worldly adventure.

There was serious discussion of scientific and philosophic matters, too. A decade ago, Pershing Square boasted a rare and natural-born folk teacher. Middle-aged, but youthful-looking, half artist and half scientist, he called himself Roger Dagan, and he was a cogent popularizer of Dewey's pragmatism and instrumentalism and of Einstein's relativity. He spoke quietly, with clarity, incisiveness, earthiness, and poetic insight. Intellectuals of all sorts, those of the schools and those outside the schools, gathered eagerly for Dagan's lectures and for questions and conversation afterward. Even the faddists, cultists, prophets, messiahs, and primitive evangelists had a respect for Dagan, and there was a legend that he had been invited by Einstein for an afternoon's conversation at Princeton.

Unquestionably, the old bughouse squares, the well-known ones and the lesser ones (for almost any city of any size has had a bughouse square of sorts), are now in decline. And the decline has been precipitate, a matter of a decade or so.

Columbus Circle as a center of bughouse activities no longer exists; these activities have been eradicated to make way for "progress."

In Union Square the crowds are smaller and much more subdued, and life has gone out of the discussions. Political debate has degenerated into political gossip, and the names of movie stars and Broadway hits are heard as frequently as the names of liberal and leftist political and labor leaders. The large mass rallies, which once attracted fifteen or twenty thousand of the faithful and the dedicated from all parts of the city to demand some reform or protest some outrage, are now passé. Isidore Wisotsky, writing wistfully about the decline of Union Square, laments that the parking meters along Seventeenth Street, where roaring thousands in the 1920s and 1930s frequently demonstrated and remonstrated, are but robot headstones marking the graveyard of a place which once was the world symbol of free speech.

New York's Washington Square has been "saved" physically, at least in a measure, if you can disregard the intimidating university buildings and tall, slick new apartment houses which enshadow it; but most of its old Bohemian life is gone. Today, a Max Bodenheim would be a curiosity in the Square, no longer indigenous as he and his kind were in the 1920s and 1930s. For Bohemia, as Robert Dunavan has described recently in *The Saturday Review,* is also in decline, and what is left of it in New York is no longer concentrated in the Village.

Bughouse Square in Chicago is also suffering the fate of Union Square in New York. Few rubbernecks now stop before the Newberry Library, even during the "season."

The decline of Pershing Square is no less real than the decline of the other bughouse squares, although the crowds still gather there, and outwardly there is much picturesque and vivid activity. The Square was "renovated" a few years ago, and parking garages were built under it, requiring the coming and going of unsympathetic citizens who use the entering and exiting escalators. The grassy interior of the park has been closed to all, thus confining the activities to the walks surrounding the park. And—is there simply no escaping it?—canned music now blares from the loudspeakers scattered around the walks. Advocates of rightist and leftist movements have all but disappeared. There is less serious debate and discussion about religious, philosophic, and scientific matters, and what there is has deteriorated in quality. Even the faddists, cultists, prophets, and messiahs have declined in numbers, variety, imagination, and exuberance.

Two things have increased: primitive religious evangelism and sheer exhibitionism. Interspersed everywhere, and at all times, there are the transplanted rural Bible-belters exhorting to "Come to Jesus" and "Get right with the Lord." And there is no end of freakish exhibitionism: the old Iowa buck dressed like an adolescent hipster; the Texas Negro got up like the end man in a minstrel show; the octogenarian Okie with long white hair tied in a knot with ribbons; an extroverted oldster from Nebraska bedecked from head to foot in jewelry, medals, and flags; an aging monstrosity from Kansas with clawing, painted fingernails six inches long. These oddities do not pray, preach, sing, dance, debate, or discuss; they merely parade. This is not vitality; it is sheer vulgarity and senility.

In their decline, all of the squares show the same characteristics, and these are usually reflections of current trends in American life.

There is much less political agitation and discussion in our public squares than there was a few decades or even one decade ago. Extreme left-wing and right-wing movements have declined in America, and since 1952 America's Democratic-Republican politics have become even more middle-of-the-road and blurred than usual. In recent years, many of the liberals among America's intellectuals have turned from political and economic issues to an examination of the apparently other-directed, non-autonomous, and conformist nature of mass man and mass society. These cultural considerations are more obscure and abstract than the older issues, and the failure of bughouse square participants to take them up illustrates a growing gap between our intellectuals and the folk. (Incidentally, the thesis that we live in a more conformist society may help explain the general decline of our bughouse squares.) Again, foreign policy is more remote and impersonal than the older domestic issues, and the realities of today's foreign policy are far more complex than the old clear-cut issue of whether America should or should not engage in world politics at all. When religion is still seriously debated, it is in terms of the conflict between the old orthodoxy and the rationalism of the Enlightenment, the rationalism translated into American folk idiom by Ingersoll and Darrow. However, our intellectuals have passed on to Freud, the subconscious, the place of the nonrational in life, the interplay of the rational and the nonrational in man. All of this has scarcely penetrated our bughouse squares—still another example of the current gap between the folk and the intellectuals. Teachers like Dagan, who attempted to explain the new psychology and the new physics, have just about disappeared.

In our county-seat towns, too, the old courthouse square is in decline. The days when these squares were centers of social activity, gossip, and community and political debate are almost gone. The benches have come to be deserted; in many places they have been taken away; increasingly courthouse grounds are being converted into parking lots. Perhaps we are already close to the day when the word "park" will connote not grass and trees and idlers but merely a place to leave an automobile. Our politicians are discovering to their dismay that Americans will no longer gather at the square for rallies, that this central characteristic of American politics for generations is now passing.

The decline of the public square probably marks a turning point in our social and cultural history. It is a part of the larger movement toward the mass society, the less personal and less neighborly society. The automobile, the drive-in, the movie, the radio, TV, and the newspaper columnist have taken the place of the park bench, the curbstone, the soapbox, the animated group debate, the Punch and Judy show, the dancing bear, the organ-grinder with his performing monkey, and the hawker of pretzels, hot tamales, sliced coconut, and candied melon rinds. Foreign immigration has slowed to a dribble, and the second and third generations of immigrant stock have become Americanized and conformist; they too have taken to the suburbs and become enamored of creature comforts. There are fewer individualists and eccentrics today, fewer fringe movements in politics and religion, fewer extremes of right and left, fewer devotees of strange and esoteric cults and heresies. Knowledge has become departmentalized, and we have become a nation of technicians and specialists. Fewer people seem interested in the larger and deeper questions of the meaning of human existence, the true ends of life—and hence these questions are discussed and debated less. Even those persons who carry on the liberal and humanist tradition have themselves become "intellectuals"—that is, specialists—and there seems to be a growing gap between them as intellectual elite and the folk.

However, we must beware of swallowing whole the thesis that we live in a completely mass society, nonautonomous and evil. It may be that many of those who formerly would have been Bohemians and habitués of the squares now find personal fulfillment in university extension and adult education classes and in today's wider scope for careers in research, the pure arts, and the proliferating applied arts.

Nor need we yield as often as we do to the encroachments on neighborliness and personality. In our city-planning, particularly in our shopping, community, and civic centers, why should we not encourage in miniature something modestly akin to those snug and cozy little squares, in the past often gems of architectural harmony—like the Piazza Navona in Rome, the Place des Vosges in Paris, Market Square in Brussels, scores of charming old squares in London, and even Mt. Vernon Place in Baltimore and

Washington Square in New York? In our current developments, we discourage graciousness and leisure, and as for a bughouse square, people could not have one even if they wanted it, for sheer lack of a converging place.

For one, as I walk along Michigan Drive or Wilshire Boulevard, stretching relentlessly for miles and miles and flanked by giant skyscrapers, hotels, and apartment houses, I feel intimidated and overwhelmed. Where can I escape for pause, for casual and informal social intercourse, for a moment of affirmation that I live in a personal and manageable world? It is as though we were back to the pre-Western epoch of history, back to the time temples like Karnak and Luxor were built to the glorification of depersonality and the belittlement of man.

✑ Letter to a New PhD

In the 1920s, when the author was a college student, he felt that professors were a rather idealistic and otherworldly lot, and widely sensitive to the broad and varied facets of Western civilization. Today, academicians are better trained but less broadly educated. They have become or are becoming specialists, technicians, careerists, and "sophisticates"—and are privileged in the worldly sense. The impact of the affluent and technocratic society on this and other aspects of the academic world is discussed informally in the piece below.

■ DEAR RIP:

And so at last you have that PhD and are in the market for a college teaching job. It is perhaps unsporting of me to recall how you, a student in my freshman class in history, feelingly protested when I predicted you would become a college professor!

Your letter expresses the usual relief in being emancipated from the tutelage of an academic committee and the joyous sense of freedom in at last being on your own intellectually. I concur heartily that the degree is merely a point of departure for the more important work and achievement that lie ahead. Yes, it is true that some of our most successful and distinguished university professors have not had advanced degrees. I like what the elder Dumas is reputed to have told his son: "A writer, my boy, never stops studying. Only professors can afford to do that. There are no degrees for us. Only performance counts." In this connection, it might be well for you to consider teaching in some of those smaller colleges of high quality which increasingly are staffing their faculties with successful poets, playwrights, novelists, literary critics, artists, musicians, actors, theatrical directors, journalists, and writers in the social sciences and the humanities who have few or no academic kudos but are actively creating in their own fields. In such an environment, one may be less likely to suffer from academic dry-rot.

Teachers College Record, LXIII, (December 1961).

You have been in the academic world long enough to realize that whatever the popular myths, colleges and universities cannot be graded like commodities. Even the oldest and most distinguished have their mediocrities, their time-servers, their wire-pullers, their beneficiaries of family pull and nepotism, their internal jealousies and conspiracies, their weaker departments, their departments rent with nasty feuds. In some of our universities with the highest prestige, there are now such pressures for publication and such intensive competition as actually to inhibit productivity in certain sensitive natures. Some people do their best creative work under pressure, but others are at their best when they have a feeling of leisure and security.

All of the desirable teaching positions are not in the East or in the big schools. Some of the most challenging and satisfying ones are in the smaller schools and in "the provinces." There is an amazing range in the kind of institutions we have in America; some of the newer, bolder, more experimental places should not be overlooked. Each school and department ought to be considered in the light of their concrete particulars and your special needs, talents, and inclinations. The lesser-known schools, however, must be given close scrutiny. You can afford to take nothing for granted. Even when the salaries, promotion policies, and tenure compare favorably with the well-known institutions, you must take a careful look at housing, insurance programs, retirement systems, teaching loads, size of classes, secretarial and typing assistance, policies on sabbaticals, library and other working facilities, and so forth.

The character of the community in which the institution is located is also important in a number of ways. Does it provide good public schools and other cultural advantages? Is it the kind of community likely to invite personal and academic freedom or the kind likely to apply social pressures to curb them? Remember, too, that the nonacademic personnel of an institution are drawn largely from the community, and these nonacademic personnel, like faculty wives, in many and subtle ways affect the atmosphere of an institution.

You may not realize it, but there are still some colleges and universities in the United States which are run not merely nominally but actually by boards of trustees or boards of control. In such an institution the lay board reduces the president and deans to puppets and actively determines educational policy. You will, of course, shun such primitive institutions as you would the plague.

Cultural Ethos and Technocracy

It is common in academic circles to belittle deans and administrators, to say that they are fixers with public-relations mentalities who went in for administration after they failed as teachers and scholars; but as you know, all of this is gross distortion. There are all kinds of college presidents and all kinds of deans. (Incidentally, have you noticed with what avidity the most acidulous of these critics ensconce themselves in administrative jobs when the opportunities arise?) It seems to me that the American system of administrative control has some advantages over the European system of faculty control, where all too often little oligarchies of established professors tyrannize over the younger men and reduce them to disciples and sycophants. In any event, since American institutions are in the hands of the administrators, you must take a long look at the kind of administrators in positions of power.

Are the president and leading deans the kind of individuals with whom you feel at home professionally and personally? Have they come out of genuinely academic and scholarly backgrounds? Are they themselves people who have been creative teachers, scholars, researchers? Or have they been drawn from outside the academic world and from the nonscholarly areas of the university? Do they have the qualifications to pass some independent judgment on your teaching and scholarly productivity, or must they depend entirely on the evaluations of subdeans and heads of departments? Do they respect honest teaching, scholarship, and research so much that they would stoutly defend them against all political, group, and public pressures? Can they talk and plan only in terms of numbers, buildings, and budgets, or can they, in addition, think and act constructively in terms of scholarly substance, educational goals, and cultural and spiritual values? Make no mistake about it: The presence or absence of an enlightened administration is the acid test of an American college or university. Avoid a university that is administered mechanically, in a bureaucratic spirit and with an eye primarily to public relations.

Next to the general administration, you will want to appraise your department. Do you respect the head of the department professionally and personally and have confidence in the way he administers? Do you respect your prospective colleagues in the department, and do you like them, or enough of them, socially? Are they of one school of thought in your discipline, and if so, do you belong to that school of thought? If not, will they tolerate dissent? Better still, is the department essentially nonsectarian, composed of members with many different approaches to the discipline? Is the department reasonably harmonious, or is it beset with strife, personal or professional? As you must know by now, some departments are bitterly divided on ideological or methodological questions, ridiculous as these may seem to a layman. If you take one side, you are damned by the other; if you attempt to stand above the conflict, you are damned by both. These feuds extend to the graduate students and sometimes even jeopardize their futures; indeed, the whole atmosphere of the department is poisoned. Steer clear of such departments.

Again, how will your work as a teacher and scholar be evaluated? There is a current and growing tendency for teachers to be rated by their students in formal questionnaires. This is devilish. It is another rod of conformity and a giant stride toward reducing teaching from art to mere technique. There is also a tendency for colleagues in a department to rate one another in formal questionnaires or for the senior professors to rate their juniors by this method. This gives free play to jealousy and all sorts of unwholesome jockeying. And some institutions—some of the better ones, too—refuse to promote to a full professorship without getting the evaluations of some professors in the same subject in other institutions. This, of course, is part of the larger trend to centralization, bureaucracy, hierarchy, and the worship of "big names." It forces those on the make to get to the national professional meetings and cultivate those who have arrived, and it gives an unfair advantage to those who went to the big graduate schools over those who did not.

All mechanical methods of rating teaching and scholarship are inhibitive and oppressive. They tend to exalt form over substance and spirit, and they stack the cards in favor of the mediocre and against the gifted. The process of determining who is a good teacher is a slow, gradual, informal, and spontaneous one. The true verdict comes out of a general consensus, developed over the years by students, colleagues, and administrators on the spot. Avoid institutions which have gone in for formalized devices of evaluation.

As a beginner, you will be expected to teach freshman and sophomore survey courses. You probably will never have a more stimulating teaching experience than this. However, survey courses differ widely. If your institution has a general education program, you will be called upon to teach not a survey of history, in which you are specialized, but a course which combines all of the social sciences; and if you had been trained in literature, you would be called on to teach a survey not of literature, but of the humanities. Now, teaching a survey in a general education program ought to be an exciting and a broadening adventure, but in actual practice it is often a letdown, even a cruel disappointment. In teaching a survey of your specialty, you are generally given a free hand. But in a general education survey, on the theory that you do not know enough about all the subjects involved, your teaching materials are usually contained in a teaching guide, prepared by older members of the staff, and you are asked to follow it. Very often all sections of the same course, taught by a great variety of instructors, are given common tests, and thus you not only lose control of the testing and even the evaluation of your students, but you also find that your teaching degenerates into preparing them for these common tests, since neither you nor your students want to fall below the ratings of the other sections. This leads to wooden teaching, and it can become mere drudgery. Again, faculty morale may be low where instructors, eager to teach the subjects in which they are specialized, are given little opportunity to do so.

In the junior colleges, general education courses may even be more unsatisfactory, for here they frequently have been oversimplified to the point of banality. Increasingly, too, instructors in junior colleges are required to have methods courses from teachers' colleges, to be certified just like public school teachers. Indeed, there is a trend toward making our community junior colleges merely grades thirteen and fourteen in the public school system.

It would, then, probably be wiser for you to choose an institution in which you will teach survey courses in your own specialty. You should recognize, however, the deep impact the general education movement has had on college teaching. Because of that movement, the survey in your specialty will be a broader and richer one, drawing much more than formerly on related fields to illuminate it. For instance, in your own subject of history, the survey you teach will be infused with materials from political science, economics, sociology, anthropology, psychology, and other related fields in a way that would have been unthinkable before the general education movement emphasized the interrelated nature of subject matter.

Many of your generation take a cavalier attitude toward the college or university in which they teach. They like to emphasize that the faculty member's first loyalty is to his discipline and not to the educational institution. This is often the excuse for showing little interest in the institution—its traditions, problems, or students. Where do those of your generation satisfy the very human need for community? After all, your discipline, even when it has a strong professional association, is largely abstract and impersonal, and most college teachers find that they cannot become integrated into the life of the town or city in the same intimate way that others who make their livings there can. Town and gown relations, even when not actually strained, are at best pretty artificial. It is to the college community that the academician must look for his genuinely close and warm human contacts, and the young instructor who looks down his nose at his institution is cutting off the social and psychological satisfactions he needs.

Many professors not only detach themselves from the college community but make no bones about being bored with teaching itself. Teaching and students are just necessary evils, tolerated only for the sake of that monthly pay check. When academicians speak of "my work," they increasingly mean their own research and writing. But such a man is really selling himself—and his discipline—short, for the investment of energy and imagination the good instructor makes in his teaching often pays fine dividends in clarifying his own thinking and generating new ideas in him. In the vital give and take of discussion and in delivering lectures, without notes, under the stimulus of listening students with alert minds, the instructor often has flashes of insight, sees relationships in new and original ways. Good teaching can be an intensely creative experience for both students and teacher, perhaps more often for the teacher than for the students.

Good teaching does not stop with the classroom. The relations of teacher and student should be personal and human, and the teacher must see inside the minds of his students. This entails meaningful personal conferences and the reading with curiosity and discernment of student papers and examinations. (The delegation of paper-reading to graduate students and assistants is a perversion of the whole teacher-student relationship.) This entails, too, faculty participation in student organizations and campus panels, forums, conferences, and discussion groups. And for the instructor who can do it naturally, it entails sitting down with groups of students in campus hangouts and, over coffee or beer, "bulling" about anything from cabbages to kings. Now, Rip, you understand that I am not exactly advising you to emulate the example of the great Galileo, who many a night made the rounds of the pubs with his advanced and admiring students and had to be put to bed by them, for I do want you to hang on to your job; but you will have the good sense to operate both humanly and prudently.

For one like myself, who has been on a college campus since the 1920s, when Billy Phelps and John Erskine were in their heyday, it is disturbing to see the increasing number of bright and brusque young careerists who each year appear on our college campuses as beginning teachers. These are immersed in "their work," eager for the main chance, busily engaged in doing the things they think will win them the approval of the foundations, the publishers, and the "big names" in their disciplines. They approach teaching with impatience and disdain any campus or outside-class association with their students. Some of these "comers" are furiously preoccupied with analyses of the organization man, the bureaucratic society, the nonautonomous individual, other-directed behavior, Madison Avenue attitudes, and the decline of personal and human values. One thing is certain: If we all operated as they do, the nightmare society they so strenuously deplore (on paper) would surely be upon us.

The question of teaching by TV is bound to become a burning one in your day. TV teaching, of course, can never substitute for personal conferences and small discussion classes, but it probably can be made as effective as our present large lectures, provided that the lecturer, *not* the professional TV technician or co-ordinator, is left with the initiative and the final decisions about content and manner of presentation. The great danger is that the technician or co-ordinator will substitute his judgment for that of the teacher. The technician will have a lot of technical and routine information about how to use charts, when to run in documentaries, etc. But the most important ideas are abstract. They cannot be pictorialized; they depend for mature understanding on mathematical symbols or skillful verbalization. The lecturer can be guided in some of the externals by the suggestions of the TV technicians and co-ordinators, but in any conflict over content and the way of presenting it, the experienced teacher must prevail. About this there can be simply no equivocation and no compromise.

You should avoid confining your social contacts to a small segment of the college community. At first you will be thrown more frequently with the members of your department, but you will soon weary of going to cocktail parties each weekend with the same people you see in Old Manse Hall every working day. I know your preference for the professional humanists, and I tend to share it, but you must have discovered by now that there are among them a disproportionate number of pedants, dogmatists, precious dilettantes, poseurs, and snobs, and a lot of mossy gentility and arid traditionalism. It may take you time to discover that delightful companions—and genuine humanists—are to be found here and there all over the campus, sometimes in the most unsuspected departments. I know, too, your prejudices against schools of education. I share them in part, because I resent both the encroachment of teachers'-college courses on subject-matter courses in the certification of public school teachers and the lobbying in state legislatures by professors of education to bring this about. It must be remembered, however, that teachers' colleges have inherited the philosophy of John Dewey; and looking back over the years, I must acknowledge that in concrete cases involving student or faculty discipline, academic freedom, desegregation, and the like, faculty members from teachers' colleges have usually been on the side of the angels. More numerously and more frequently than the professional humanists, the educationists have stood for humanity and liberty.

As to your nonteaching professional career, join the national and regional associations of your discipline, attend what meetings you can (at your university's expense), and don't affect disdain for the programs. There are always some good offerings, and it is a valuable experience for young men like you to observe in action some of the older heads from all over the country. However, don't become an association politician with an ambition to hold the offices; the time and energy spent in such a fashion can be more fruitfully employed.

Intellectually, examine the trends and fashions which happen to be current at any given time among your professional colleagues, but never surrender your independent judgment. Specialize in those aspects of the subject you like best, interpret in the light of your own insights, and use those methods which seem to you most productive.

Don't publish just to get published. Almost anybody can publish, for the learned and professional journals here and abroad are numerous and proliferating all the time. True, deans in some schools will not know the difference between a creative article and a stale one or between a journal of prestige and a peripheral one. With the increasing pressures to publish or perish, it is fortunate for run-of-the-mill academicians that there are so many outlets and so few discerning deans. But I am assuming that you will be in a place where administrators are more aware. Besides, for your own respect and the respect of those you respect, you will be careful about what and where you publish. Incidentally, take a crack now and then at the weekly, monthly, and quarterly journals of opinion and literature. This will require you to improve your style, give you wider scope for interpretive ideas and judgments, allow you to appear in the company of writers and thinkers outside the academic world, and boost your morale occasionally by providing you with a little extra cash earned in the practical competition of the market place.

You say you want to publish a trade book with literary as well as professional merit. By all means work toward this objective, but don't be too much in a hurry. You still have some romantic notions about the publishing business. There are few of the old, leisurely family firms operating today. Many of the commercial houses are now big businesses whose stock is sold on the exchanges. You mention certain "preferred" publishers, and I can see that you have neatly rated them, just as you patly classified colleges and universities before you knew more about them. Actually, there are dozens of first-class commercial houses, and the university presses are growing in quality and prestige. In dealing with your publisher, much of your satisfaction (or the lack of it) will depend on the editor to whom you are assigned. Whether the editor you get is gifted and sensitive or merely a pretentious nonentity will, alas, be largely a matter of chance. If you write a good book, it will probably be recognized as such, even though you publish with one of the university presses or "lesser" commercial houses, and even though it is not reviewed in "the big places." Some books conspicuously reviewed are remaindered within a few months; some that are quite inadequately reviewed survive to become influential.

You take a dim view of textbooks. True, textbooks vary widely in quality. Some are bad, and the majority are mediocre. But a considerable number are highly meritorious, and a very few are distinguished enough to conquer the trade market and even become paperback classics. In the case of most textbooks, the returns to the author are usually quite modest; but if you write a widely adopted book, your royalties will be larger than those from a best-selling nonfiction trade book, and they may even exceed those of a best-selling novel by a literary celebrity.

In all probability, you will not be teaching long before you are approached by publishers to do a textbook. When this happens, you likely will be surprised, for nine times out of ten the subject about which you will be asked to write will already have been covered by scores of up-to-date texts. The profits from this side of the business are great and the competition among the publishers keen. It is a pretty safe bet, Rip, that your present indifference to writing a textbook will be overcome by the prospect of the big money and the expectation that your textbook will be one of the exceptional few to rise above the conventional market. If you should feel the urge to write a text, my advice to you is this: Wait until your middle years of maturity, scrutinize your contract well, and then write a manuscript which covers the basic material but, in addition, is crammed with insights and ideas that will differentiate it from the run-of-the-mill variety.

Of late, the term "academic prostitute" has come into increasing use. This has reference to the growing number of professorial rackets. Even though you have been living in a university environment for a number of years, I am afraid you have not been enough of an insider to recognize their nature and extent. There are many of these, and I shall mention only a few.

For instance, should you become chairman of a leading survey course in a big university, you may have as many as five-thousand students enrolled in your course during the school year. Publishers will sight a real bonanza—a captive market of five-thousand students each year—and you will be importuned to write a text for the course. Believe it or not, there are professors so rapacious that they will consent to do this and thus exact a personal tax from their students. Occasionally, a text written under these conditions is so shoddy that the publisher confines its sale to that university alone and withholds it from the general market, an indication of how profitable a single captive market can be. Again, there are professors who get leaves of absence on foundation grants, ostensibly to improve their teaching by observing in other universities or to enrich their experience by travel or research, and actually use this time to write textbooks, and sometimes grade-school textbooks at that. There are resident research professors, too, who use the time allowed them for "research" to write textbooks.

As you are well aware, there are now scores and scores of foundation and government grants for professional travel, research, and lecturing in foreign universities. Some academicians have now become highly peripatetic; they are away from their home universities on various grants as often as they are in residence. Their home universities have become mere bases from which to operate. In many instances, too, the habitual beneficiaries of grants have few or no publications to show for their privileged experiences. Rip, this is the kind of racket you are most likely to fall into, because you have your PhD from one of the big, prestigious universities, many of whose faculty members sit on the national boards which select the recipients of these grants or, in any event, carry great weight when they make recommendations to board members. It has been my observation that a few of the "big name" professors have been rather shameless in the way they have gathered the plums for their former graduate students, occasionally a great variety of plums for the same individual.

The rackets thus far mentioned are the fancy ones. There are others, more common garden varieties, particularly in the colleges and universities in the provinces. For instance, some faculty members give only the most perfunctory performances in their regular academic jobs and the better part of their time and energies to personal enterprises of a non-academic kind—speculating in real estate, administering their own rental properties, running cattle or dairy farms, conducting private consulting businesses, and so forth. These practices are rare among the penurious instructors, much more common among the older and more affluent full professors.

The most pervasive academic corruptions, however, are of a less tangible and more tragic kind: the gradual oozing away of youthful enthusiasm and idealism; the bureaucratization and impersonalization of procedures; the everlasting angling for place and promotion; the search for a soft and secure life without practical or intellectual competition; the overemphasis on specialization, minutiae, and trivia; and the refusal to look to the larger consequences of one's work, its place in the larger whole. There are the pressures to conformity and gentility; the exaggerated concern for what administrators, colleagues, and even the public will think; the fear to teach and write honestly and creatively and the animosity toward those who do; the timid hesitation to be a genuine and spontaneous human being even in one's personal life. Many will say, "But these are the corruptions of life itself, particularly in our kind of industrial society." And I reply: "We yield the good life more often than we need to, and if individuality, spontaneity, and creativity cannot be encouraged in the colleges and universities, are not these institutions abdicating some of their chief reasons for being?"

My wish for you, Rip, is that you will escape not only the grosser forms of academic corruption but also the less tangible and more subtle ones, and that in being true to the instincts which brought you into the academic world in the first place, you will achieve those wonderful and abiding satisfactions that come from imaginative teaching, original research, and creative thinking.

The big foundations are playing a larger part in the academic world. In applying for a research grant, don't trim your project to suit what you think are the foundation's preferences. Stick to the subject you believe to be important and to your own methods for doing it. It may be necessary, however, to do a little camouflaging. You will have to formulate a hypothesis and a "model" of the procedures you propose to use and to conform to a certain jargon in describing these. There is a game that has to be played, and you must play it with a straight face. Once you get the grant, it is quite permissible to alter your hypothesis and modify your methods in the light of the greater knowledge and experience your intensive research provides. Indeed, to do otherwise would be "unscientific." Rip, this is not dishonest; it is merely being pragmatic, getting the hang of how to wangle a grant in order to do the honest work in the subject you think significant and by the methods you think valid.

Some day, in mid-life, you may become a dean! Don't dismiss the idea out of hand. Some deans manage to function as capable administrators and still salvage their scholarly productivity. If all intellectuals turned their backs on administration, how would we have satisfactory working conditions for the faculties? As for taking an executive job with one of the foundations, that is another matter. It is possible, but it is not likely, that your creativity would survive in that rarefied atmosphere.

What a change has come over the academic world since I entered it over thirty-five years ago! How quiet, unworldly, and innocent it was then, how modest the material surroundings and rewards, how few the opportunities for glamorous careers. College professors were a kindlier, more unpretentious, more humane, more earthy, less "sophisticated" breed. The homespun personalities typified by David Starr Jordan, John Dewey, Charles A. Beard, and John R. Commons were in every educational institution. A college campus was certainly no Eden, as illustrated by the fate that befell Beard, but even so, there were fewer pitfalls.

But after all these years, Rip, I still love to teach. I am still charged with a positive thrill when I enter the classroom, and at the end of the period, I often say to myself, "And to think I get paid for doing this!" With all its drawbacks, college teaching probably still has fewer irritations, anxieties, pressures, and corrupting influences than the other ways by which men make a living. In providing leisure and opportunities for creative activity, a modern university is superior to a Renaissance patron. Have I, then, at times seemed hypercritical? Compared with the anguished outcries of a da Vinci or a Cellini at a capricious princeling benefactor, I have, with reason, been mild indeed.

And now a final word. If you are a successful and popular teacher and a productive scholar, you are likely to arouse the jealousy, malice, and conspiratorial instincts of your less gifted and less energetic colleagues.

Cultural Ethos and Technocracy

Remember, Rip, with all its relative advantages and gracious amenities, you are entering what has now become one of the bitchiest professions in the world, and don't let your guard down.

Every good wish for a successful career with a minimum of compromise and for as happy a life as is vouchsafed us mortals.

Sincerely yours,
Bill

❦ American Intellectuals
and American Democracy

The traditional American society has been pragmatic, inventive, and dynamic. Its class lines have been fluid. There has been an impulse to improve the lot of the underdog, a long-time drive to egalitarianism. That society has also been pluralistic and diverse, as evidenced by the various origins of its population and by the amazingly wide range and complexity of its economic life, which has led to broad latitudes of choice and pronounced libertarian characteristics. The American society, too, has not been without folk quality on its intellectual, aesthetic, and spiritual side. These attributes have been emphasized in the following article

The other side of the American society—its drive to distortion in the direction of the practical and the material; the relative weakness of Western civilization's tradition of humanism and spirit of *joie de vivre*; and contemporary technocracy's apparent trend to generate as many conformist and nonautonomous tendencies as pluralistic and libertarian ones, giving rise to certain Orwellian threats—will be treated in the succeeding article, the one entitled "The Century of Technocracy."

Both articles should be read together; they complement each other. The American society today is so portentous and complicated that all generalizations must be made cautiously. For almost every trend there is a counter-trend, and within that society there are groups which defy every generalization about the American culture made in the following article and the succeeding one.

Running through the article below is a tone of irritation, for the author feels strongly that more of us must look at our society comprehensively, difficult as that is, if we are to take our bearings, discern undesirable trends, and make intelligent attempts to correct them. But, as this article ventures to point out, American intellectuals of the 1940s and 1950s, instead of making the comprehensive approach, too often retreated into minutiae, intensive specialization, overconcern for methodology, an exquisite preoccupation with the individual psyche and individual complexes and neuroses, an avoidance of commitment, or a renunciation of all values except a faith in scientific techniques;

Antioch Review, XIX (December 1959)

or fell back on abstractionism, dialectic, and a new scholasticism in search of absolutes, eternal verities, and a monistic tradition.

One who attempts a comprehensive examination of the nature and quality of our complicated American society today, which is in frenetic transition from the old to the new, blazing a trail to a strange civilization, must be content merely to throw out ideas and suggestions rather than draw definitive conclusions. However, because the seeker after the large view cannot yet come to sure conclusions about what kind of society is emerging in America, it does not follow that he should surrender his own values; rather he should exert himself to make that emerging society the Good Society, in the light of those values—always, though, within the framework of historical realities and possibilities, never in terms of historical make-believe or utopia.

■ THE WORK and values of American intellectuals of the 1920s and 1930s have been under critical fire since World War II. But what of the intellectuals of the 1940s and 1950s? Perhaps by now it is not unduly presumptuous to begin attempting to put the American intellectuals of mid-century in some kind of historical perspective.

First, a word about the intellectuals of the 1920s and 1930s. The present generation tends too much to regard American thinkers and creative artists of those years as "radical." True, many of them were excited by the constructive possibilities of the Russian Revolution and enormously interested in the Soviet Union as a laboratory for vast social experimentation. However, as Granville Hicks and others have repeatedly pointed out, very few of America's intellectuals of the 1930s were Communists. Even among those who went pretty far to the left, there were many who survived their earlier enthusiasm to make valuable contributions to democratic thinking in the 1940s and 1950s. Some of these—Edmund Wilson, Malcolm Cowley, Robert Gorham Davis, Hicks himself, and others—reveal an awareness of twentieth-century realities, born of their earlier probings and experiences, too often lacking in our younger contributors. Others, with the God-that-failed mentality, having become disillusioned with the authoritarian "truth" of the left, have been searching ever since for some brand of authoritarian "truth" on the right. These embittered futilitarians, often extroverts in their personalities, have been among the leading assailants of their own generation.

It is my own feeling that the intellectual temper prior to World War II was more in harmony with the dominant American tradition than is the intellectual temper today; and more, that in general the intellectuals of the earlier period were coming to grips more realistically with the central problems, domestic and international, of our time.

Much has been written about how we were "betrayed" by the intellectuals of the 1930s. But as we move deeper into the twentieth century, we may find that it was the intellectuals of the 1940s and 1950s, and not those of the 1930s, who departed more widely from the American tradition—at the very time, too, when that tradition needed to be reinterpreted and applied to the realities of today.

II

What is the dominant American tradition? It is optimistic, democratic, rational, experimental, and pragmatic. It maintains that men are not the slaves of social conditions and blind historical forces, that to a large extent men can rationally mold their own institutions, without a hereditary or a privileged elite to guide them, and within the value-framework of human dignity and individual freedom. Although this tradition is often rationalized in terms of Locke, the Enlightenment, and Rousseauan humanitarianism, it came out of the historical interplay of creative man and indigenous American conditions: vast land and resources and a relatively small population; the relative absence of feudal, manorial, communal village, guild, and internal mercantilistic practices and traditions; the lack of aristocracy, priesthood, status, and "orders" generally; frontier realities; and the obviously howling success of the American experience.

America's Declaration of Independence boldly proclaimed the equality of men, regardless of birth, status, race, or religion. It was a revolutionary document when proclaimed, and it is a revolutionary document today. The Constitution of 1787, "struck off by the mind of man," created a government structure so contrived as to guarantee liberty; it set an example of how man might literally "make" his own institutions. Even the Transcendentalists, in revolt against rationalism, were optimistic and practical idealists. Like the older American rationalists and the later American pragmatists, they were ready to make a future to order. "Why should we not also enjoy an original relation to the universe?" asked Emerson. As H. S. Commager has pointed out, the Transcendentalists, with all their faith in a priori truths, took a chance that the heart knew better than the head and labored heroically to make the good come true. Even the most lasting impression of Darwinism on the American mind was not the rationalizations of the Social Darwinians glorifying the plutocrats as the "survival of

237

the fittest," but the assurance that there is constant change; and change, to most Americans, was something that could be directed to the good, be made "progressive." And at the turn of the century came William James's pragmatism, that marriage of American optimism and practicality with the bright promise of science. Pagmatism harmonized with American tradition and practice, for it emphasized the importance of experimentation, breaking with the past, rejecting custom and habit, trying new methods, creating a future on order.

American intellectuals of the 1920s and 1930s were in the dominant American tradition. John Dewey emphasized the plasticity of man's instincts, the malleability of human nature. Truth was not only what worked out for the individual but for society. Individuals sought, in common with their fellow men, for secular, immediate, and particular truth that had meaning for the community. Thorstein Veblen, fusing economics, sociology, and anthropology, and posing as an objective, if ironic, observer, sought through an analysis of the conflicts between business and industry, between technology and the price system, to realize the possibilities of the new technology. Charles A. Beard and Herbert Croly saw in the marriage of Hamiltonian centralization and Jeffersonian democracy solutions of contemporary industrial problems through democratic national planning. Vernon L. Parrington wrote passionately of the history of American thought as the history of an evolving revolutionary liberalism; the central theme was that of the battle between conservatism and liberalism, reaction and revolt, with liberalism and revolt triumphant; the American tradition emerged as the identification of Americanism with democracy. Roscoe Pound, Oliver Wendell Holmes Jr., Louis Brandeis, and Benjamin Cardozo were in revolt against legal abstractionism and scholasticism, and were busy infusing the law with pragmatic values and social realism. Sinclair Lewis was writing of the ugliness of commercial civilization; Dos Passos and Steinbeck and Farrell were portraying social injustices; but in all of these writers, even in the determinism of Dreiser and Darrow, there was the unspoken assumption that man could overcome ugliness and evil by changing his environment. Even Thomas Wolfe, at the close of his life, was groping toward a recognition of something larger than himself.

In contrast, the most vocal and conspicuous of the present generation of American intellectuals have surrendered to noninvolvement or noncommitment, or retired into formalism, or become obsessed with techniques, or retreated into the individual psyche, or fled into the nonrational and the irrational, or seriously distorted the American democratic tradition, or sought to substitute for evolving and fluid democratic values the fixed values of status and aristocracy. Exponents and exemplars of each of these points of view have all played down the wondrously rich diversity and flexibility of American life.

Now, of course, the older trends have not disappeared; the dominant points of view of the 1920s and 1930s still have creative exponents today; and vistas opened by earlier thinkers continue to be explored. In some areas, notably law and jurisprudence, pragmatism and social realism continue to make notable gains. Also, it must not be supposed that the new conservative points of view are entirely new; all of them had antecedents in the 1920s and 1930s and even earlier. Nor can it be denied that some of the work of America's uncommitted and conservative thinkers and creators of today is deepening insights, sharpening tools and techniques, penetrating psychic mysteries and complexities, intensifying aesthetic perceptions, and in some aspects of thought and life making Americans a more discriminating and sophisticated people.

What is novel in the current situation is the popularity of noninvolvement and noncommitment and of conservative and aristocratic values. Conservatism is being proclaimed as the truly traditional attitude of Americans. Conservatism, it is said, has been our dominant tradition all the time; what is new, it is claimed, is our belated recognition of conservatism as our national tradition.

III

Many intellectuals of the postwar generation have fled from all values whatsoever. They have taken refuge in noncommitment and noninvolvement and called it objectivity or sophistication or wisdom. The great vogue among textbook writers and publishers for anthologies and collaborative works which acquaint the student with "all points of view" is today a much-used way of escaping integration and commitment. Sampling opinion, taking polls, and compiling the results of interviews constitute another way of escaping commitment by merely tabulating statistically the relative opinions and values of others on comparatively surface or safe questions. More and more, authors of books and magazine articles merely report opinion but do not express opinions of their own. They indicate trends but seldom pass judgment on them. And as for moral judgments—these are to be avoided as "hortatory" and "evangelistic." There is not even a commitment to eclecticism, to an avowed defense of a society in which the existence of plural and diverse values is taken as a positive good, the hallmark of a healthy, vital, free, and infinitely fertile civilization. This would be a defensible position, for it may be that the American society has in fact become such an eclectic society. But few contemporary intellectuals are committed to eclecticism as a positive value in itself.

Closely akin to the trend to noncommitment is the emphasis on form, technique, and methodology. One way of escaping commitment in subject matter is to concentrate on form. In all the intellectual disciplines today, even in the humanities and the social sciences, there is a growing concentration on forging methods and techniques that will make the discipline truly "scientific." Now, of course, nobody objects to sharpening the tools of investigation and research, but an excessive concern for methodology may also be a way of avoiding all substantive import. For instance, had Kinsey waited to "perfect" his methodology, or retreated into a minuscule area of investigation, or confined himself to a safe aspect of his subject, or been frightened by the moral and social implications of his researches, we would have been denied one of the few illuminating works of our time.

In literature, too, there has been an almost obsessive preoccupation with form, with style. If the 1920s and 1930s represented a revolt from formalism, the present generation is characterized by a return to formalism. T. S. Eliot and the New Critics began the trend to form; today, form has become the vogue and most of our influential writers have been molded by the canons of the New Criticism. Norman Podhoretz has commented on the high sytlistic polish, the precociously sophisticated craftsmanship, of our rising young novelists, even when these have little to say.

In matters of substance, many of this generation have retreated from social considerations into the individual psyche. There is an enormous preoccupation with individual man's motivations, love life, sex drive, frustrations, complexes, and neuroses; and there is a neglect, even an ignoring, not only of social considerations and problems but even of the impact of society and environment on individual man himself. Podhoretz has also commented on the concentration of today's young novelists on the individual's psychological drama. To our rising novelists, the supreme fact is personality and the main business of life is love. There is an almost total lack of awareness of the social environment as a molder of character and as a maker of the traumatic situations themselves.

In even many of our mature novelists and playwrights, life has been reduced to the individual's viscera, his gonads, and an eternal contemplation of his navel. There is, of course, a certain insight and fascination in the plight of the neurotic hero enmeshed in his own complexes and in the interplay of his little circle of family, friends, loves, and hates. But is this the ultimate wisdom in art? On this question, Arthur Miller eloquently writes in the New York *Times*:

What moves us in art is becoming a narrower and narrower esthetic fragment of life. . . . The documentation of man's loneliness is not in itself and for itself ultimate wisdom. . . . Analytical psychology, when so intensely exploited as to reduce the world to the size of a man's abdomen and his fate equated with his neurosis, is a re-emergence of romanticism. It is inclined to deny all other forces until man is only a complex. It presupposes an autonomy in the human character that, in a word, is false. A neurosis is not a fate but an effect.

Is all of this an escape? Miller suggests strongly that it is, that we actually are more aware than previous generations of the impact of the city, the nation, the world, and now of the universe on our individual lives; yet we persist in refusing to face the consequences of this in art—and in life.[1]

The degree to which nonsocial and even antisocial attitudes have come to prevail in literature and humanities faculties of our colleges and universities today is perhaps realized only by those of us who teach in them. Recently, by way of illustration, it was argued in the humanities staff of one of our state universities that Steinbeck's *Grapes of Wrath*, which for many years had been required reading in a course in twentieth-century literature, should be dropped from the readings because it was "only a social tract and not literature." The book was dropped.

Historians and biographers, whether they have the training for it or not, increasingly use the psychoanalytic approach in writing the lives of famous leaders. This method is also used by Harold Lasswell and his followers to analyze the careers of prominent politicians. When this is the

---

1. The self-indulgent and inbred quality of much American writing was all the more striking when compared with the post–World War II literature produced in Europe at the same time. In Europe, the impact of Man's tragedy at this juncture of history was everywhere deeply felt. Many German writers were busy expiating the sins of the Nazis and warning of the smoldering social hates persisting in Germany. Boris Pasternak literally gave his life that he might tell the story of personal trauma caught in the wider trauma of history. A host of Polish, Italian, and French writers, in various ways, were expressing with burning eloquence the piercing historical epic of our time. Some of these, like the haunting Quasimodo, were dismissed by Americans as mere left-wingers. But who can deny the authenticity of the humanist tradition at its best in Albert Camus? The awful tides of twentieth-century history seem to course through Camus's own soul to give us the social and spiritual insights of *The Myth of Sisyphus* and of *The Rebel*.

approach, the tendency is to emphasize personal motivations, often running back into childhood, and not the social consequences of the leader's mature activities. Since conservatives are less "'troubled" than liberals, radicals, and rebels, they usually come off as more adjusted and normal, whereas the nonconservatives, being more "troubled," are likely to come off as agitators, extroverts, and cranks, as victims of frustrations, complexes, and neuroses. This method, which minimizes the larger social setting of a leader's work and the social results of his work, tends to reduce all leadership, conservative and liberal alike, to the trivial and the commonplace, to rob it of its historical and social significance.

During the 1920s and 1930s, advances had been made in treating the leading philosophers and political and social thinkers in terms of the historical time and social milieu in which they did their thinking. Many intellectuals of this generation are in revolt from this. Exponents of the "great books" approach, the Neo-Thomists, and, among others, David Easton in political theory argue that this method reduces thought to a mere sociology of knowledge. There is now a new emphasis on "pure thought," on theories separated from their historical and social context, on sheer logical analysis of the ideas themselves in an attempt to discover absolute truths. However, the emphasis does not seem to be on all the acknowledged ranking thinkers in the Western tradition. All too often greater emphasis seems to be placed on the thinkers in the Plato-Aquinas tradition, in the a priori-deductive-authoritarian tradition, and less emphasis placed on the equally important thinkers in the skeptic-nominalist-inductive-empirical-pragmatic tradition. Now, intellectuals engaged in this business are certainly not running from commitment. On the contrary, these are seeking, through a new exegesis and a new scholasticism, absolute commitment and eternal verity.

IV

Many of today's intellectuals are putting a new emphasis on nonrational values. The rationalist values of the Enlightenment are more and more under attack. This is particularly true among the New Critics, the Southern Agrarian Romantics, and the increasingly influential Neo-Thomists. Arnold Toynbee, who has a greater vogue in the United States than anywhere else, speaks disparagingly of the Enlightenment as turning its back on the Christian virtues of faith, hope, and charity, and emphazing the Mephistophelian maladies of disillusionment, apprehension, and cynicism. A political scientist, Eric Voegelin, is writing a gargantuan work, an attempt to build "order" in history and politics on Thomist thought. It seems that the apple of discord appeared with the Greek skeptics and Medieval Gnostics.

Now, of course, rationalism is not enough to explain history and life, but neither is nonrationalism. In order to be aware of the importance of the nonrational, one need not embrace nonrationalism completely; indeed, a rational respect for nonrational values may be the beginning of wisdom. Our society today, conscious of both, has a better chance of reconciling rational and nonrational values—of balancing the Apollonian and the Dionysian (to use Ruth Benedict's terms), the prudential and the passionate (to use Bertrand Russell's terms)—than any society which ever existed. But this new balance and this new wisdom will not be attained by denunciation or rejection of rational values.[2]

Closely connected with the revolt from rationalism is the revolt from democracy and the bold affirmation of the values of an aristocratic society. The intellectual revolt from democracy had its origins in the 1920s with H. L. Mencken, the New Humanism of Irving Babbitt and Paul Elmer More, Southern Agrarian Romanticism, and the New Criticism. Ortega y Gasset's *The Revolt of the Masses*, which appeared in 1930, has had a tremendous, though often unacknowledged, impact on nondemocratic and antidemocratic thinking in America. Today, Southern Agrarianism lives on in John Crowe Ransom, Allen Tate, and a number of younger votaries. The Neo-Thomism of Mortimer Adler has increasing influence. And the New Criticism of T. S. Eliot, Ezra Pound, and Kenneth Burke has become the dominant influence in literary criticism. "America," according to Burke, "is the purest concentration point of the vices and the vulgarities of the world." The New Criticism has been summarized by Robert Gorham Davis as a way of thinking in which "authority, hierarchy, catholicism, aristocracy, tradition, absolutes, dogmas, and truths become related terms of honor while liberalism, naturalism, scien-

2. The Americans remain a prudential people; despite today's outward appearances they continue to be suspicious of nonrational behavior in both its destructive and its highly constructive sides. As for the "better chance" Americans have for reconciling the Apollonian and the Dionysian, this is true in the sense that Americans have greater wealth and leisure to indulge their emotions and passions in amusements and pleasures, that Freud has become more popularized in America than elsewhere, that psychiatry and psychiatrists are more in vogue, and that more writers portray the play of the nonrational in the individual psyche; but all this is countered by the ramifications and disguises of a persisting puritanism and the relative frailty throughout American history and life of humanism and the *joie-de-vivre* spirit, to be discussed in the next article.

tism, individualism, equalitarianism, progress, protestantism, pragmatism, and personality become related terms of rejection and contempt."

In his *The Conservative Mind*, Russell Kirk finds in conservatism a unified movement and a consistent body of first principles from the time of Edmund Burke right down to the present. He sees these first principles as belief that a divine intent rules society; affection for the mystery of traditional life; conviction that society requires orders, classes, status, social gradations, and hierarchy; persuasion that property and freedom are inseparably connected; faith in prescription; recognition that change and reform are not identical. This, of course, is a far better description of traditional European conservatism than of American tradition or even of American business "conservatism."

Peter Viereck, in his *Conservatism Revisited*, finds in Metternich and in the Metternich system of 1815–1848 a sagacious attempt to conserve the traditional values of Western civilization and to bridle the forces of liberalism, nationalism, and democracy, which, according to Viereck, have led in the twentieth century to the mass man, fierce class and international wars, and totalitarian statism. A defense of the Metternich system is something novel in American historians, for up to this time every school of American history—Federalist, Whig, and Democratic—has seen in the Metternich system mostly obscurantist reaction. Many American specialists in international relations are searching in the Metternich system and the diplomacy of the reactionary Castlereagh for techniques and methods for coping with the revolutionary ferment of our time.

There is a disposition, too, to import from anthropology the "tenacity of the mores" as reason for resisting change. "Respect for the mores" played its part in the decision to retain the Mikado in Japan; it is being employed to sabotage new racial adjustments in the South. A century, it seems, is hardly long enough to allow Southern whites to adjust to the changes of the Civil War. Anthropology, of course, deals with the ways and techniques of change as well as with the resistances to change, but the emphasis of the conservative intellectuals who go to anthropology for rationales is on resistance to change.

America's New Conservatives describe the European past in too-glowing terms; they romanticize it; they hearken back to a Golden Age which in fact never existed. They minimize the good of an industrial society. They do not give enough credit to the enormous gains made by industrialism over poverty, ignorance, disease, and personal brutality. They make conservatism both too comprehensive and too simple. They tend to appropriate for conservatism the whole humanistic tradition of the Western world, whereas, of course, that tradition has become a part, in somewhat varying ways, of Western reaction, conservatism, liberalism, radicalism, and socialism. Again, a consciousness of tradition, habit, and the organic continuities of social life is not so much a matter of conservatism as of the maturity and profundity of one's social understanding. The New Conservatives see in historic conservatism a consistency, a unity, and a continuity it never had in fact. If the need today is to weld aristocrats, businessmen, Catholics, and Protestants into a common front against Marxism, such was not the case in the past; indeed, such a combination represents a putting together of historic enemies. The contemporaries of Metternich did not consider him an impartial mediator between aristocrats and bourgeoisie; they knew him to be the friend of aristocracy and the enemy of the bourgeoisie. Very often these New Conservatives think of conservatism as being above the battle, as being the wise judge composing conflicting interests, snugly fitting moderate change into social and historical continuity. They too often forget that conservatives, like everybody else, are motivated by their interests. Conservatives are likely to overestimate the virtues and underestimate the injustices of an existing order and of their own position. As William Lee Miller has pointed out, "relative justice" will usually not be found with the conservatives, however cultured or humane or "new," but with their opponents.

The New Conservatism is an attempt to substitute European conservative values for the American liberal tradition. It is as transparent a fraud as the Southern antebellum feudal dream of Fitzhugh, Harper, Ruffin, Holmes, Hughes, Dew, DeBow, Tucker, Bledsoe, and Hammond. (How many Americans, how many Southerners, remember these names today?) Kirk and his cohorts are attempting to foist Burke's traditionalism on America, and as Louis Hartz has acutely pointed out, Americans, including Southerners, cannot become Burkean traditionalists without becoming Lockeans, because the predominant tradition Americans have had is that of Locke.

V

Some other thinkers are guilty not of falsifying the American tradition but of distorting it. Frederick Hayek, who has had an influence on contemporary American thought, erects free enterprise into a rigid system and decries all departures from it as "the road to serfdom." Walter Lippmann, in his *The Public Philosophy*, takes natural-law concepts, which are capable of both a liberal and a conservative interpretation, and gives them a conservative Thomist slant. (Lippmann himself is an illustration of many of the older generation who have traveled spiritually from left to right. Lippmann has moved from democratic socialism to liberalism to conservatism—and to the urbane observer above the battle.) Louis Hartz, Clinton Rossiter, and Daniel J. Boorstin emphasize the private-property drives of Americans to the point of distortion.[3]

Americans have believed that they began in revolution and that by a continuous process of experience and free experiment have been in a continuous revolution, a "permanent revolution," ever since. Hartz gives American history a different twist. He contends that Americans have never had a revolution, that they were born free. Unlike the European, they escaped the feudal-manorial-guild society, the society of aristocracy, status, and fixed orders, and were literally born into the new society of the free market. The result was a property-owning, middle-class society which took Locke for its patron saint; and if Locke had never existed, this society would have had to invent him. Now, of course, there is much truth in this, but it overstates the ease with which the American society came to birth and maturity and it understates the difficulties of building and maintaining a free and democratic society in the face of constantly recurring tendencies in the human situation to privilege, complacency, stratification, and ossification. In spite of the relative absence of an inhibiting cake of custom from the past and in spite of the favorable conditions of the American environment, a free society did not just occur automatically. Even in America, inhibiting leftovers of the European past were considerable. To realize the possibilities of the new environment, to mold America's free society and adapt it to constantly changing conditions, required vision, imagination, and humanitarian impulse; it required the experimental and pragmatic spirit; it required struggle and the willingness to do battle; it required innovators and fighters and innovating and fighting movements.

3. This reference to Hartz, Rossiter, and Boorstin is misleading. These differ from one another in their emphasis. Besides, one can scarcely go wrong in laying stress on the compulsive acquisitiveness of Americans. The real failing of most of those who play up the American urge to material well-being and the "practical" is their neglect to give sufficient attention to the traditional American drives to humanitarian reform, social uplift, moral crusade, and an ever-widening egalitarianism, even allowing for the obvious racism running through American life and history. Americans have been better humanitarians than humanists.

Kirk claims that the American Revolution was "a conservative restoration." Viereck calls it a "conservative revolution" in the pattern of the Revolution of 1688. (This would indeed be a surprise to Macaulay, who described the American experiment as "all sail and no ballast.") But Hartz says it was no revolution at all. (How "hot-house" all this would have seemed to the expropriated and exiled Tories!) The American Revolution, with its Declaration of Independence and its sweeping away of Old World legal, political, economic, and social "vestiges," *was* a revolution, a revolution of considerable radical propensities.

Even though most of the subsequent political and social drives in American history represented "merely" a further breaking down of the status society, an extension of the free market, and a more widespread distribution of private property, they also showed a continuing humanitarian and innovating spirit. The great mass movements to abolish indentured servitude and imprisonment for debt, to enfranchise the propertyless, to emancipate women, to educate the masses at public expense, to abolish slavery, and to insure social justice in an industrial society are not to be set down as merely putting the finishing touches on a system preordained by indigenous American conditions. The struggles between theocracy and Independency, Old World tyranny and New World freedom, seaboard and frontier, federalism and republicanism, agrarianism and capitalism, slavery and freedom, industry and labor were hardly picayune. Even the party battles of Hamilton and Jefferson, Clay and Jackson, McKinley and Bryan were not the hollow shams a whole host of young historians and political scientists are now "demonstrating" them to have been. While the drive in America to private property, to "American Whiggery," and to Horatio Algerism, so emphasized by Hartz, is undeniable, so also have been the drives to the freedom of the mind, equality of opportunity, and relative social justice—as exemplified in varied ways by Roger Williams, Daniel Shays, Franklin, Paine, Jefferson, Benjamin Rush, Jackson, Dorothea Dix, Horace Mann, Emerson, Greeley, Theodore Parker, Ellery Channing, Wendell Phillips, Lincoln, Whitman, Peter Cooper, Henry George, Edward Bellamy, Peter Altgeld, Susan Anthony, Jane Addams, George Norris, John Dewey, and scores of others.

Hartz raises some basic questions about the very nature of the political process. Are the conflicts of politics, even in Europe, rendered meaningful only when they involve class and the dialectics and ideology of class? Cannot problem-solving and the clashes of group interests, as distinct from those of class, also be meaningful? Granted that American politics have involved little class and ideological conflict, have not the clashes of America's amazingly diverse group interests over the distribution of the benefits of the American economy been most significant? Was the New Deal any less significant because its approach was nonideological, that of "mere" problem-solving? Cannot Americans make decisive contributions to the underdeveloped peoples today by a problem-solving rather than a dialectical-ideological approach?

The Hartz thesis, that America has had no feudal and aristocratic right, no Marxist left, but only a liberal center, does much to clarify the difficulty Americans are having in understanding the "isms" abroad. But in playing down the humanitarian and the experimental elements in American life, Hartz and other intellectuals are making it more difficult to bridge the spiritual gap between Americans and the non-Communist social-democratic revolutionary forces abroad. We cannot bridge this gap by harping on America's "monolithic liberalism," but we may bridge it by emphasizing the innovating, pragmatic, and basically nondoctrinaire nature of Americans and American society.

VI

Those Americans writing about the American economy in a large and significant way—and beginning with World War II these have dwindled in number—are guilty of one kind of distortion or another.

A. A. Berle, who with Gardiner Means in the 1930s wrote a most meaningful book about America's big corporations, has now become an apologist for the economic concentrates. In his *The Twentieth Century Capitalist Revolution*, he likens the big corporations to the feudatories of old and to the modern sovereign state. Indeed, Berle sees the modern corporation taking over some of the functions of the state. According to him, we need not fear this corporate power because the big corporations will check one another, and they will go a long way toward checking themselves because of the sense of benevolence and social responsibility they are developing, the self-restraints of natural law, and the infusing spirit of a kind of twentieth-century City of God. Berle's distortion is that of optimism.

C. Wright Mills, in his *The Power Elite*, sees big business and the political state as merged, and the United States as a monolithic oligarchy ruled by the corporate rich, the high political directorate, and the war lords. These orders of the American society have an interlocking membership and transfer among them takes place at the top. But what is the common goal of this elite? What common interest holds its parts together and gives it a common direction? Mills develops no concept of class which might hold this elite together, and he suggests no other bond of unity. What, then, prevents rifts and conflicts in the elite itself? The truth seems to be that the American society is more complex and diverse and has more vital conflicts and clashes on all levels than Mills will concede. America's elite is not as exclusive, concentrated, and unified as he makes out, and its decisions are not as free from the influence of Congress, political leaders, political parties, group associations and pressures, mass organizations, journalistic opinion, and public opinion generally as he contends. Mill's distortion is that of oversimplification and pessimism.

John K. Galbraith, until recently, strongly suggested that we may enjoy the economic benefits of oligopoly and at the same time escape the evils of concentrated power because of the operation of what he calls countervailing power, new restraints on economic power which have come into existence to take the place of the old competition. The new restraints appear not on the same side of the market but on the opposite side, not with competitors but with customers and suppliers. Galbraith admits that under the inflationary pressures of demand, countervailing power weakens and then dissolves. But even in the absence of inflationary pressures, countervailing power is not pervasive but sporadic, because while in our economy some customers and suppliers are large-scale and organized and able to exert countervailing power against original power, others are small-scale and unorganized and unable to exert countervailing power. Galbraith suggests that government step in and create countervailing power where bargainers are small-scale and unorganized. But government is not something isolated from society, above the battle, forever standing ready to act in a disinterested way to correct imbalances. Government itself is subject to pressures. Why doesn't government encourage the countervailing power of small business to oppose the original power of the concentrates? Because the concentrates have access to government, too; they can put pressures on government; indeed, their pressures are likely to be more effective than those of small business. Gal.-braith's distortion is that of seeing too much symmetry and balance, of substituting countervailing power for the old "unseen hand" as a built-in regulator, as an automatic balancer.

However, one can scarcely classify Galbraith as a conservative. In his more recent *The Affluent Society*, he has revealed some skepticism about his own concept of countervailing power, and he has emphasized the importance of the public sector, the area of government-produced goods and services, in our economy. Moreover, he promises to analyze the core of our economy in a forthcoming book on the corporation and the corporate personality.

Is it not strange that in the avalanche of publications that pours from the American presses there are so few studies of our business concentrates, those huge institutions that lie at the heart of our economy and our society? There is no end of studies in the classical market tradition, but there are relatively few which deal institutionally and pragmatically with our big-business economy. Every year our universities and foundations sponsor all sorts of studies on threadbare and peripheral subjects in all the intellectual disciplines, but how few studies there are that deal with the actual operation and administration of our giant corporations. Just how independent are these corporations? What is the nature and effectiveness of their controls over the economy? Why have not the trained personnel in these big corporations, more and more of whom are specialists from our graduate schools, taken advantage of their positions inside these corporations to give us realistic studies of their actual workings? (At least after they have retired or gone on to other employment.) Until more of these studies are made and then synthesized, much of what we say about the American economy—and the scope and nature of power in America today—will be inadequate, unrealistic, and speculative. Impressionistically, our economy seems to be amazingly mixed; but whether there is a preponderance of power and where it is, or what are the more important loci of power and whether and how they are balanced, are questions which cannot be answered yet.

VII

Perhaps the most pervasive distortion today is the belief that we live in a completely mass society and mass culture and that these are evil. Many intellectuals and aesthetes, and their numerous imitators, speak rather glibly of mass man, organization man, the crowd mind, group conformity, "togetherness," nonautonomous man, stereotyped man, the other-directed society, and the monolithic society. This conception may be selling our liberal society short. On the other hand, this may be symptomatic of a growing awareness of the importance of cultural freedom, in addition to political and economic freedom, a manifestation of a cultural revolution which is already far along the way. People were not so conscious of agrarianism's narrow limitations and its nonautonomy, but at least we are now increasingly conscious of the nonautonomy produced by industrialism.

The truth seems to be that while in some of its aspects our society is more integrated than the preindustrial societies of the past, in some other aspects it is less integrated. In its economic and political areas it is more integrated, and there are threats to personal autonomy in these areas. However, we are increasingly conscious of them, and it is precisely in the political and economic areas that the American liberal tradition is strongest. This liberal tradition needs to be reinterpreted and then applied to the new realities. Certainly, rejecting the liberal tradition at this time would be a surrender to the forces of impersonality and not a challenge to them.

But in many of its other facets our society is less integrated. People in an industrial society have been freed from the old restraints of localism, from customary status and class, from the traditional primary groups of family, neighborhood, and parish. They are mobile. They can escape to the anonymity of the cities. There is a much wider range of choice in careers than there was in the past. And all people, no matter what they do, work less and less and have more and more time on their own. But do they escape the old fetters only to be absorbed in the crowd? There is escape into privacy and into subgroups of one's own choosing as well as absorption into the crowd.

W. H. Auden, in his *The Dyer's Hand*, observes that advanced technological society, by putting at our immediate disposal the arts of all ages and cultures, has completely changed the meaning of the word "tradition." It no longer means a way handed down from one generation to the next. It means a consciousness of the whole of the past of all societies— with their infinite ways and values—*as present*.

In our society, all sorts of values—preindustrial and industrial, rural and urban, nonscientific and scientific, nonrational and rational—compete and jostle within communities, neighborhoods, families, and even individuals. (In some individuals, this produces neuroses; in many others, it produces personal emancipation and cultural enrichment.) In their personal lives, Americans display a wide variety of recreational, aesthetic, religious, moral, and sex values—perhaps the widest variety in the history of human societies. The amazing variety of attitudes and practices in sex found by Kinsey would likely be found also in other aspects of personal living, if these were given similar scrutiny.

The mass media, which are often said to impair autonomy and creativity, are in fact Janus-faced. They widen the outlets for banality and mass vulgarity, but they also widen the outlets for good reading, good drama, good art, good music, informed interpretation of events, and live history-in-the-making. Actually, in consumption, there seems to be at the present time more discriminating connoisseurship in food, drink, clothing, dress, housing, sex, entertainment, travel, reading, music, and the arts than there was in the days before mass advertising and the other mass media. And in production, creativity, far from drying up, has never been so alive. Never before has it been so easy for so many people to have careers in research, scholarship, science, and the pure arts. Since 1901, American scientists have in each successive decade been receiving a larger and larger portion of the Nobel prizes. Since the 1920s, Americans have had five winners of the Nobel prize for literature. We may deplore the current avoidance of social realism by our contemporary American writers, but we can only applaud their increasing sophistication and craftsmanship. And the practical arts grow and multiply. There has been an enormous expansion and proliferation of the service industries, of applied social science and psychology (like social and personnel work and psychiatry), of applied writing (like public relations, trade journalism, criticism), of applied art (like architecture, landscape gardening, commercial art, dress design).

A self-confessed middlebrow, Russell Lynes, comes closer than any intellectual to summing up the nature of our culture:

Ours is a "You-name-it-we-have-it" kind of culture. It is a vast market place of conflicting tastes, conflicting ambitions, and conflicting needs. In guaranteeing "the pursuit of happiness," we recognize that not every man's happiness is measured by the same yardstick. We may do our damnedest to convince him that our yardstick is better than his, but we do not beat him over the head with it. . . . Out of the crowd that the *voyeurs* of culture call "the mass," many single voices are heard. So long as this is true, what we have is not a "mass culture," but neither is it an aristocratic culture. It is a highly competitive culture.[4]

Is not ours a society without a metaphysical base? Is not ours an amazing diverse, pluralistic, and eclectic society held together by a kind of humanistic pragmatism? And is not ours the probable prototype for other advanced industrial societies of the future?[5]

4. "Proof that We Are Not Barbarians," *The New York Times Magazine*, July 6. 1958.
5. In many ways the American society has been diverse, pluralistic, and even eclectic, but the humanist tradition has been weak. In using the term "humanistic

To discover the actual nature of our society, the individual must for the most part look to the realities about him and trust to his own observations of life. Our researchers are currently bogged in methodology, minutiae, timidity, and bureaucracy; and our humanistic and literary intellectuals, with a passion for monistic unity, symmetry, conceptualism, and abstraction (most of all the New Humanists, the Neo-Thomists, the New Critics, and the New Conservatives), are busy constructing some homogeneous pattern for our society and culture, a pattern which simply does not exist. In a recent issue of *Commentary*, Robert Gorham Davis has pointed out how during this century American literary critics, with thorough Judeo-Puritan-Brahmin bookishness, have given the whole American literary tradition a homogeneity which in fact it does not possess.[6]

---

pragmatism," the author was too much influenced by what he would like to see rather than the fact as it is. What makes the American society hold together as a going concern is primarily its marvelous capacity to produce goods and services and provide wide popular access to them. A marriage of pragmatism and humanism, discussed in the next article, would doubtless give the American society a more satisfactory spiritual base and a higher-quality civilization. As for our society's being a prototype for other advanced industrial societies of the future, this is undoubtedly true in the technological sense, and most other societies, Communist and non-Communist, are striving to achieve the American standard of living. But the Communists decry many other aspects of American civilization; Western Europeans want to hold on to as much of their humanist tradition as they can; and in the underdeveloped countries there is an aspiration to combine much of America's machine technology with indigenous, preindustrial folkways and traditions. However, as machine technology spreads, other countries are likely to undergo a general process of "Americanization."

6. This article as initially published contained an additional section, but this has been eliminated because it deals with the impact of contemporary American thought on foreign policy and the reaction of foreigners to it, much of which has already been treated in other articles in this collection.

❧ The Century of Technocracy

The preceding article characterized the American society as essentially plural, pragmatic, and experimental, with a pronounced libertarian, humanitarian, and egalitarian bent. The following article considers today's growth of technocratic organizations and values and asks whether our society is not now giving rise more to nonautonomous and other-directed trends than it is to pluralistic and libertarian ones.

The article also probes more deeply into the quality of the American culture. While holding to the view of the preceding article that the American culture is far from the "mass" culture in the invidious sense in which that term is usually used and that there is vitality in our "you-name-it-we-have-it" kind of culture, it suggests that unless considerable of the folk bring to the market place high-quality values, we cannot achieve a genuinely high-quality civilization. It further suggests that an important source of high-quality values is to be found in Western civilization's tradition of humanism—in its theoretical thought, its aesthetic perceptivities, its free spirit of the *joie de vivre*. It points out that humanism has always been relatively weak in the United States and gives the reasons for this, and it argues that a revival of humanist values would go a long way to counter the inbred dehumanizing and depersonalizing tendencies of technocracy.

What is really proposed here is an enrichment of American pragmatism with Europe's traditional humanism. This would be no new American importation; it would be a revival and an enlargement of what had been inherited long ago. In so far as it has a philosophic basis, does not pragmatism itself have its roots in the nominalist-inductive-empirical side of Western civilization's secular humanism? Did not the philosophers of the Enlightenment themselves go back to the thinkers of Greece and Rome for their inspiration and insights? Was not the thought of Thomas Jefferson a combination of pragmatism and humanitarianism—and Western civilization's secular humanism in all its breadth and comprehensiveness?

Americans pioneered the technocratic society, and its roots reach far back into American history. Industrialism came first to England, but for a variety of reasons a "know-how" with tools, and later machinery, permeated the whole American population in a way that was unique.

The Antioch Review, XXV (Winter 1965–66).

It would be a tragedy if the material blessings of the technocratic civilization should be offset, perhaps more than offset, by a distortion in the direction of "soullessness"—insufficiently relieved by the high-quality intellectual, aesthetic, and spiritual values and the liberating sense of personal vivacity and joy to be found in our own Western civilization's humanist heritage—and we thus missed our chance to achieve the Good Society. As religious values decline, it becomes all the more important that humanist values be encouraged.

Although the earlier portions of the following essay repeat some materials found elsewhere in this collection, they have been retained to comprehend in a single piece the impact of technocracy on both America and the world at large.

■ WE NOW KNOW that the overriding contribution of the twentieth century to history is technocracy: the technicalized, cybernetic, computerized society increasingly run by scientists, engineers, and technicians. This society is drastically changing man's institutions and transforming his behavior and basic values.

The technicalized society and its technocratic organization are, of course, a continuation of the older industrial revolution, and they represent the cumulative impact of decades and even centuries of industrial and scientific development. But this century differs from the nineteenth in the accelerating pace of change, the omnivorous ramifications of machine practices and techniques, and the palpable influence of these on every aspect of life and thought.

Metropolitanism and machine technology are spreading to all parts of the world. Societies still differ markedly in their standards of living, their economic systems, their political practices, and their social ideologies. But the time is approaching when life, at least externally, will appear very much the same everywhere. In every society, the central reality is coming to be the ubiquity of the machine, the deference for quantitative methods, and the rule of the technicians.

The term "technocracy" has an unpleasant ring. The avowed "technocrats" of the 1930s seemed brash, extravagant in their predictions of machine affluence, cold and inhuman in their emphasis on efficiency procedures, and undemocratic in their frank acceptance of the rule of engineers and technicians. Nevertheless, the societies now emerging seem amazingly like those anticipated by the technocrats, and perceptively adumbrated by Thorstein Veblen even earlier.

255

II

The belief persists that our century has witnessed the triumph of socialism; but the Western societies today are not like those envisaged by the pre-1914 Socialists. These would have gone beyond Keynesianism and the welfare state to build a society run by dedicated collectivists, largely dependent on the democratic decisions of free labor unions and co-operatives, in which natural resources and basic industries were publicly owned and production was for use and not for private profit. Also, to the pre-World War I Socialists, heirs to the humanist tradition of the old Europe, leisure was no "problem"; rather it was the spiritual as well as material emancipation for working people which opened the way to a joyous and creative participation in cultural pursuits.

Again, the European Socialists of the early years of the twentieth century, while anticipating some of the threats of a bureaucratic society and giving thought to means by which these might be counteracted, did not foresee the astonishing degree to which the industrial society would become pervasively technicalized; the marked reduction in the working force through automation; the prominence in administration of apolitical, nonideological technicians; the ways in which humanist and qualitative values would be menaced by materialist and quantitative ones; and the Orwellian overtones in technicalized societies, the subtle inhibitions against which the traditional and formal civil liberties offer inadequate protection.

The extreme and "orthodox" Marxists would be surprised at the way Marxism has actually developed in Communist countries. During World War I and the years immediately following it, the failure of the Communists to win any country outside Russia early demonstrated that there was to be no international proletarian society, that Communist economies would be built, if built at all, on national patterns and by national states. Under Stalin, Soviet Communism became more and more nationalist, and unmistakably state capitalism rather than socialism. The labor unions, collectives, and co-operatives had no real voice in decision-making, and they became mere arms of the state. Russia emerged as an autocratic managerial society in which the decisions were made by centralized administrators, engineers, and technicians. The Soviet Union became not an example of a socialist society but a model of how a preindustrial or semi-industrial people, by submitting to stern totalitarian economics and concentrating on heavy industry at the expense of consumer wants, might amass state-owned capital and achieve an industrial revolution in a hurry.

Following World War II, when some other countries, in one way or another, were captured by the Communists, these also built their economies on national patterns and became additional examples of state capitalisms and autocratic managerial societies.

Also following World War II, nationalist anti-imperialist revolutions swept the colonial areas of Asia, the Middle East, and Africa; and scores of new nations were born. Contrary to Marxist-Leninist expectations, these anti-imperialist revolutions did not merge with Communist revolutions. In the more than twenty years since the end of World War II, little North Vietnam is the only former European colony to be captured by the Communists. (China, while economically exploited by the European powers and divided into economic spheres of influence, was never a political colony of Europe.) The new nations realized that they could raise living standards only through the adoption of machine technology, and they hoped that mixed economies (with both public and private sectors) and economic aid from advanced industrial countries would be less painful ways of achieving this than the drastic methods of totalitarian economics.

In brief, the chief instrument for improving living standards was machine technology and not any particular ideological system; and there was a variety of ways machine technology might be introduced into the underdeveloped countries. By the late 1950s many of these countries were receiving economic aid, machinery, and technicians from both the Soviet Union and the Western powers. Pragmatic, mixed economies and various versions of the administrative, managerial state were emerging.

All of this is not to say that some new nations, failing to achieve stability and rising living standards, may not still experience Communist revolutions. But it appears that if additional Communist revolutions come, they will be sporadic—that is, in one country here and in one country there; that they will not be linked to any monolithic Communist center, for no such center exists; and that each will more or less go its own national way. Nor is there any reason to suppose that any additional Communist revolutions will develop into anything other than state capitalisms and managerial societies. However, the managerial state in Communist countries is more centralized and drastic in its planning and administration than in non-Communist countries; but as Communist countries grow older, they tend to liberalize, as the Soviet Union is doing.

It is conceivable, of course, that ideological conflict will be given a new lease on life. American foreign policy might be "taken in" by the rightist myth of a monolithic, ever-expanding Communism and pursue courses which would literally force all the competing brands of Communism to unite on ideological grounds. Or American economic aid might degenerate into programs of virtual bribery seeking to keep unpopular regimes in power, regimes which do not mean business about introducing basic technological improvements. Or there might be a collapse of a number of underdeveloped countries because of population explosions and famine conditions, inviting China, still in the stage of revolutionary fervor, to intensify a Communist drive and even convert it into an antiwhite crusade.

However, such possibilities are not probabilities. The long-time historical trend is away from ideological conflict; the "savior" is increasingly seen to be machine technology and not ideology; there are a number of combinations by which underdeveloped countries, with time, may work a technological revolution in their transport, communications, agriculture, and extractive industries; and a few of them, with longer time, develop into industrial societies.

There have been repeated predictions that Spain and Portugal will experience leftist revolutions following the deaths of Franco and Salazar. But it now appears that these creaking, semi-medieval societies will escape revolutions. Their industrial capital and machine technology are slowly developing under even right-wing dictatorships; standards of living are gradually rising; and both leftists and rightists seem to be moving to a new, apolitical, nonideological materialism, accommodating themselves to a slow drift to a more mechanized society. Should this indeed be the evolution in these countries, it would furnish additional and striking evidence of the primacy of machine technology over ideology, of the great variety of ways in which such a technology takes root today.

III

The most spectacular exploits of our technocratic age are the explorations of outer space and the nuclear revolution in war.

Just what consequences man's conquest of outer space will have are at the present time not clear. That this conquest will result in a stupendous increase in our knowledge of the universe is unquestionable. But whether it will have the revolutionary economic, social, and political impact of the age of discovery and exploration in the late fifteenth and the sixteenth century, when man for the first time traversed the utmost parts of the earth, is still in doubt.

Some of the far-reaching consequences of the nuclear revolution in war, however, are already fairly clear. That revolution is eliminating war, banishing force from international affairs, ushering in the *pax atomica*. Machine technology had already rendered World War I and World War II "total." The nuclear revolution has now rendered war "totally total," and therefore obsolete. (Now, it is possible, though not probable, that man will experiment with full-scale nuclear war just once; after that, although men would survive, we may be sure that there would never be another war.)

There is still brinkmanship, but not war. The United States and the Soviet Union recoiled at Berlin and in Cuba; China recoiled at Quemoy and Matsu and in the Himalayas. The powers avoid conventional ground wars for fear these will escalate into general nuclear war. They fear even the "limited" wars which were fought in Korea and Indochina in the early 1950s, again in fear of escalation into nuclear war. The only "safe" form of force seems to be primitive guerrilla wars, and care is taken to prevent such wars from becoming "limited" wars which in turn might become nuclear ones. Guerrilla war has been going on in South Vietnam for years, yet even now, although it has been intensified, that war has not developed into "limited" war on the Korean scale.[1]

More than any other institution, war has been the most significant instrument of change in human affairs—creating new nations, destroying old nations, altering boundaries, stimulating and accelerating innovations within the nations themselves, making internal revolutions. It is difficult to conceive of history without war, and the full impact of its epochal elimination will not be felt for decades.

1. This was written in 1965, before the Vietnam war had been escalated to Korean proportions and beyond. American policy-makers, who had expected to keep the war below the Korean level, were led astray by the computer and the computer-type mind, which did not take into sufficient account the intangibles, such as the history, culture, and spirit of the Vietnamese. The American intervention, like most superpower intervention today, was to maintain the status quo. The unexpected length and scale of this "little war," instead of causing the Americans to refrain from future interventions to maintain a status quo favorable to themselves, is more likely to prompt them to alter their methods of managing these "little wars," as discussed in the Introduction to "Our Post-Crisis World." Since the nuclear balance has already made the big powers cautious about intervening directly to *change* the status quo, and since guerrillas alone, without outside help, rarely make successful revolutions, as discussed in the next article, internal revolutions will still be more difficult to achieve than they were in the prenuclear past, when the powers felt freer to intervene in behalf of change.

However, the nuclear revolution has already made many historical predictions and analogies obsolete. A third world war is now most unlikely. It now seems certain that the Communist revolution will have no Bonaparte. The prophecies of the cyclical historians have been rendered false, for our "time of troubles" will not degenerate into numerous and widespread wars, and a new "universal imperium" cannot be put together by a twentieth-century conquering Caesarism. A multiple balance of power in the traditional sense, based on counterbalancing alliances, cannot really revive, for, as Hans Morgenthau has pointed out, alliances can now never have what stability and reliability they formerly had; fear of nuclear destruction will make every member of a combination look sharply to its own survival.

The immediate effect of the Atomic Peace is the freezing of the international status quo. The actual power situations in the world seem currently sealed. Even when the balancing of power runs through countries rather than between them—as in partitioned Germany, Korea, and Vietnam—the situations cannot be altered by force and have not yielded to peaceful methods of change. The avowed purpose of American foreign policy is to prevent any territorial changes by force. American policy in Vietnam exemplifies this, and the United States has ruled out force as a means of unifying Germany and Korea. Revolutions and counterrevolutions within countries are more difficult, too, for there is fear of any significant shift in the balance of power by way of internal changes, particularly if there is interference by outside powers.

The most obvious problem posed by the Nuclear Peace is how to unfreeze the status quo, how to get necessary and desirable changes—the kind of changes formerly made by force, by internal revolutions and international wars. However, if the drive to machine technology has indeed become the central movement of this century's history, and if mechanized societies, however established and under whatever labels, succeed in satisfying the material aspirations of more and more peoples, then ideological and even national conflicts will recede, and the thawing of a frozen status quo, both internally and internationally, while still important, may in general be considerably less so in future decades than it is today, for the number of acute crises will diminish.

However, undue optimism must be guarded against. Although rising living standards and an acceptable status quo are increasingly seen to depend on the spread of machine technology, the establishment of effective machine technology may turn on: (1) native governments which have sufficient discipline and determination to work an economic transformation, and this often requires social revolution of one kind or another and charismatic leadership; (2) aid to such effective native governments by advanced industrial powers, and certainly a refraining from sabotaging them merely because they don't have the "correct" ideological label; and (3) international arms-control agreements and the consequent reduction in the military paraphernalia of the super "warfare states," thus releasing capital and technicians for constructive economic developments.

IV

The material blessings of the technicalized society are self-evident. But such a society will not realize its full potential for good, and indeed it may be pregnant with much evil, unless it is capable of controlling and mastering the machine and providing widely shared cultural and spiritual satisfactions. It would be folly to ignore the perceptive insights of Burckhardt and Nietzsche, Brooks and Henry Adams, Ortega and Orwell, Spengler and Toynbee, even when we do not accept such insights as prophecies.

Different combinations of technology produce different results in society. Are we to exercise no control at all over the rate of technological change, over the technological priorities and combinations, over the torrent of innovation, the blind drive to technocracy? Are we to be guided solely by the principle that if a new invention is technically an improvement, we will accept it at once and automatically, regardless of the social and aesthetic consequences?

Specifically, are we to exercise no voice in the rate of automation? Are we to repeat over and again such instances as committing a substanial part of our resources and scientific brains to landing a few men on the moon, with virtually no public discussion? Are we tamely to submit to the building of a national computer network which will keep an eye on the private conduct of every individual in the land?

Now, this is no plea for a latter-day Luddite movement; and emphatically, it does not suggest that scientific inquiry be fettered in any way. As Nigel Calder, editor of *New Scientist*, London, points out, a distinction must be drawn between scientific inquiry and the use made of scientific knowledge so acquired. We can regret nuclear weapons without regretting the knowledge which made them possible.

In a recent article in the *Nation*, Calder asks that the choices involved in society's applying scientific knowledge and making use of technological inventions be subjected to more conscious and deliberate decision-making, and that where they concern government they be submitted to the democratic process and widely discussed. It is ironical that our political campaigns still ignore such issues, although these affect all of us more intimately than most of the issues we do discuss.

Are such decisions really amenable to democratic decision? The whole trend of the technocratic society is away from vital decision-making by electoral, party, and parliamentary processes and toward decision-making by administrators flanked by specialists and technicians. Technological questions are complex, the very kind thought most to be in the province of "the experts." Americans are so biased in favor of technological "progress," no matter the social and aesthetic implications, that they may not always be counted on to make discriminating choices, even if choices were submitted to them. The administrators, too, cannot escape the American cultural ethos; indeed, these are increasingly drawn from the very specialists steeped in quantitative methods and values. But beyond all this is the intrinsic difficulty of the questions themselves. After all, just where does one draw the line between encouraging the computer which expedites business and research and discouraging the computer which operates as "big brother" and records the most private behavior of the citizens?

Vexatious as these questions are, we must hopefully attempt to decide the most pertinent of them through our democratic process. If democracy is to function in our technicalized society, technological issues must be translated into party terms and philosophies and thus made intelligible and meaningful to voters. The party oriented to certain social beliefs would encourage or discourage one set of technological items; the party oriented to other social beliefs would encourage or discourage another set. In this way the dilemmas and uncertainties surrounding the application of technological changes may be exposed and their complex social implications clarified. Technology is simply too important to be left entirely to the technocrats.

We have been repeatedly warned—by Riesman, Whyte, Packard, and many others—of the insidious effects of big organization, bureaucracy, and the mass media in producing other-directed behavior, the nonautonomous man, and the embryonic Orwellian society. These should never be minimized, but neither should they be exaggerated. Over a century ago, Tocqueville saw the pressures of mass democracy toward individual conformity; and before that, in the old aristocratic societies, there were not only the restraints of economic scarcity but also those of class, social custom, folk tabu, and the parochial community. The new inhibitions of the technicalized society are of a more complex kind; but at least we are more aware of our restraints than past generations were of theirs; and while many of ours today are subtle and difficult to isolate and treat, others are overt and concrete and relatively easy to come to grips with.

Certain developing practices need not be tolerated. For instance, both business and government are prying too much into the private lives of their employees and prospective employees; they are coming to know all too much about them; and if a halt is not called to this, we shall wind up an intimidated, spiritless, spineless, noncreative society. Drastic measures can and must be taken to curb the overuse and overevaluation of psychological tests and the proliferation of lie-detectors, bugging devices, and other electronic snoopers.

We must be vigilant about other new sanctions used to enforce conformity. For instance, a single additional example: Just as the theocratic and pervasively religious society called in the priest to brand nonconformity as "sin," so the technocratic society tends to call in the social scientist and psychiatrist to brand it as "sick.' Like so many other things in our contemporary society, psychiatry has a great potential for both good and evil, and its use as an enforcer of conformist behavior must be closely guarded against if we are to remain a free people.

V

A new elite is arising in America. This elite consists of strategically placed scientists and administrators in the leading universities, the national foundations, the big corporations, and government. It tends to inbreeding, centralization, and an overestimation of quantitative values. More and more of its members, including those in the top ranks of the big corporations, are the products of the graduate schools of the country's prestigious universities.

It is more difficult than formerly to reach the top, to become a ranking member in the professions, to become a decision-making executive, administrator, engineer, and specialist. To arrive at such places today, one must have a long and formal education, increasingly competitive, and pass batteries of intelligence, psychological, and diagnostic tests along the way. Those who do not "make the grade" are increasingly regarded as intellectually deficient. We seem to be headed for new caste societies, for hierarchies based on brains and training, for stratifications more invidious than those of the old aristocratic societies, where failure "to belong" did not brand one with innate intellectual inferiority.

However, many of those who fail to reach the top will not in fact be second rate. For the technocratic society is putting a premium on one kind of intelligence and markedly depreciating another kind. The tests put the emphasis on quantitative skills, for these are what are wanted, and the very tangibility of such skills makes them easily "testable." Artistic and literary imagination and social, political, and historical insight are being severely played down, in part because of the bias in favor of "the objective" and in larger part because of the intrinsic difficulty of testing such intangibles.

In a recently published essay on education remarkable for its candor, Christopher Jencks observes that all secondary and undergraduate college education is coming to be dominated by the standards of the graduate schools, that in the future the college teacher primarily interested in teaching undergraduates will be "either an eccentric or an incompetent." Apparently the practice of teachers and undergraduates thrashing out ideas and conflicting values is coming to an end. The whole bent is toward an unquestioning acceptance of technocratic values; and we are moving to even more emphasis on methodology, tools, techniques, expertise, and to an even more minute subdivision of research. Increasingly the attitude seems to be: "If it can't be quantified, don't bother." If this trend is not tempered, almost everything will depend on facility in quantitative methodology: getting into college; being admitted to graduate school, no matter the area of specialization; winning scholarships and fellowships; achieving postdoctoral opportunities; securing desirable positions; being promoted to the top echelons.

Under the impact of quantification and the abstractionism to which it leads, our very language is becoming less concrete and metaphorical and more impersonal and abstract. Robert Cluett, writing in *Teachers College Record*, observes: "The principal actors in our cosmic drama of technology and bureaucracy are trends, factors, machines. . . . After speaking too long of a reality in which man does not count, we might find ourselves in 30 years ready to yield up our world to the machine and abstract forces that we have heretofore been able, in our clumsy way, to resist and sometimes control." The "other realities" which do not lend themselves readily to quantification—the worlds of perception and insight, of analogy and anomaly, of the personal and idiosyncratic, of the fortuitous and unexpected, of the unclassified and unpredictable, of the paradoxical and ironical—may wither in man's consciousness along with the very language formerly used to describe them. A Shakespeare scholar recently said to me: "I have no assurance at all that future generations will have an appreciation of either Shakespeare's dramatic situations or of his language."

We condemn the Communists for rejecting much of the rich humanistic heritage of Western civilization. But is not technocratic America repudiating much of the substantive thought and methods of thinking of the West's humanist tradition? Humanism in all its aspects has always been much weaker in America than in Europe, and now it is being further weakened.

From colonial times to the present, discerning foreign travelers and commentators have noted in the American society a "missing ingredient": the failure of American intellectuals to immerse themselves in the humanist writings and thought of the ancient world and Europe (the best of nearly three thousand years of Western civilization), thus depriving educated Americans and American elites of the canons for making critical judgments of comparative philosophies and values and the materials for thinking in theoretical terms, even theoretical scientific terms. André Siegfried, writing in the 1920s, was struck by the way advanced industrialism and machine technology were moving Americans still further away from humanist values, and he wondered whether the old European civilization had really crossed the Atlantic at all.

In America, to be sure, Greek and Latin were taught in the academies and colleges, but the emphasis was on grammar and syntax ("mental discipline") and on the gathering of ornamental quotations for public address, and not on the spirit of secular humanism and the content and thought of the authors. Even the humanism of the Judeo-Christian tradition was somewhat weakened in America because of the strength of those Protestant sects least hospitable to humanism, and because, in frontier and remote rural areas, religion frequently degenerated into primitive evangelism. True, in our large cities and universities, there have always been a few individuals thoroughly at home in the best literature and thought of Western civilization; but these have never been representative of American society or of any sizable group or class within it; and today, while the number of specialized scholars in some branches of the humanities is increasing, the number of broad humanists is declining.[2]

The essays of Montaigne are examples of how for centuries European humanists have used classical thinking for stimulating their own insights, and of how revealing of human behavior, individual and social, these insights are. For instance, after reading Montaigne's "On Some Lines from Virgil"—which is an exposition of the human sex drive, its many psychic and social ramifications, and the essential similarity of male and female behavior—one wonders whether the Freudians have explored these matters any more discerningly than the French essayist of the sixteenth century.

A recently published book, *Human Behavior*, by Berelson and Steiner, catalogues over one thousand generalizations about human behavior, based, according to Stuart Chase, "not on intuition, theology, the wisdom of the ages, or the standard belief systems of the tribes, but on scientific research." Now, it is all very well to boast of the scientific "discoveries" listed in this book, but what is most impressive about it is the large number of its generalizations that were anticipated by the humanists. Moreover, there is a vast number of generalizations contributed by the humanists not yet touched by this book or by science; the humanist generalizations cover a much wider gamut of life than do the generalizations in the Berelson book; and most of them have relevance for both industrial and preindustrial societies, whereas most of the generalizations

2. In order to find a powerful class in America interested in supporting humanism, one must go back to the leaders of the New England Puritans of the seventeenth century, who modeled their university and grammar school curricula on the Renaissance curricula of Cambridge University and the English Latin grammar schools, and who instilled in ordinary folk some respect for humanist education. Their influence, although diluted through time, was later transmitted to other areas of the United States by way of transplanted New Englanders. It is a commentary on the weakness of humanism in America that one of its most important sources is found in the Puritan elite, whose chief interest, after all, was religious, and whose humanism was less impregnated with aesthetic sensitivity and the *joie de vivre* than was that of Europe. With the relative decline of Congregationalism and Anglicanism and the growth of the evangelical Protestant sects, humanism became further diluted. That branch of secular humanism known as the Enlightenment had a pervasive influence on educated Americans, but in this case there was no class sufficiently interested and powerful to impress the intellectual and theoretical aspects of Enlightenment thinking on the American folk mind. That bundle of American folk values often attributed to the Enlightenment—pragmatism, democracy, egalitarianism—came largely out of the indigenous American environment. The folk took to Jefferson's politics, but not to his humanist "free thinking" or his classical and Renaissance sense of beauty.

in Berelson have relevance for only industrial societies. However, Berelson contains some generalizations not found in wide-ranging Montaigne, and I daresay not suggested anywhere in humanist literature; and scientific method has certified many of the insights of humanist knowledge, made them more exact and convincing, more applicable and usable.

Have we forgotten that all of the sciences had their origins mainly in philosophy, theology, and secular humanism? Have we forgotten, too, that as late as the nineteenth century and the early years of the twentieth century, humanist studies were throwing out their seminal suggestions to scientists steeped in humanist knowledge? Marx, one of the founders of social science, was enormously influenced by Hegel. Freud was obviously much stimulated by the myths, folklore, and legends of early man and classical times. Einstein has acknowledged a debt to Kant.

Even in Europe, humanism is increasingly on the defensive as "Americanization" and technicalization accelerate their pace. It is imperative that we take measures to restore and invigorate humanist knowledge and thought. This means using not only the "big names" like Plato, Aristotle, and Aquinas, but also scores and scores of humanist writers and thinkers who today are well-nigh forgotten, even by our "humanists." These should be read, in the original or in translation, for content, substance, ideas, insights. Such a restoration is necessary to give wider meaning to life, balance the now-distorted bent toward science, and encourage the type of broad historical and humanistic thinking in public affairs represented by Walter Lippmann. (Without such a restoration the Lippmanns will soon be an extinct breed, leaving us only the behaviorists, game-theorists, and computer-minded.) But more, a humanist revival is necessary to vitalize science itself, keep it from ossification, suggest new areas for scientific investigation, widen the frontiers of scientific inquiry, stimulate new scientific breakthroughs.

Historians as far removed in basic outlook as Gordon Childe and Arnold Toynbee, both surveying the rise and fall of civilizations, are agreed on this: that whenever in the past, technology has become concentrated and centralized, exclusive in its methods, and inbred in its values and elites, the resulting distortion has caused that technology to lose momentum and then to atrophy.

This suggestion of a broad humanist revival is neither quixotic nor novel. Modern technology, increasing wealth, and the accumulated scholarship of centuries have given us access today to a knowledge of cultures on an unprecedented scale. W. H. Auden has commented that works of the cultures of societies of all ages are now so widely and readily available as to make their art and values ours—*not as past but as present*. Our task is to put this in both comprehensive and manageable forms, to be used at the various levels of the educational process, so that more and more Americans may become aware of the wide latitudes and choices open to them and increasingly able to make the comparative cultural evaluations that will lead to new and seminal departures in the arts and in the sciences.

VI

The drive to economic production and individual practical success is still dominant in American life; and the tensions in "getting ahead" increase as technology steps up the tempo of the economic world, techniques proliferate and rapidly change, and it becomes more difficult to keep "on top" of one's job. But if the tensions increase, the time spent at work decreases. However, instead of enjoying the additional leisure, more and more people "moonlight," seek supplementary and odd jobs. Americans face retirement more reluctantly than do people in other countries. Many Americans dread it; they delay retirement as long as possible; they gratefully go on working in postretirement jobs; and when retirement overtakes them at last, they suddenly discover they have no inner resources to tap, no spiritual cravings to satisfy.

But despite the gospel of work, the need for human labor is diminishing. People will go to school longer; they will retire earlier; even when in the economic harness, the hours at work will progressively decline. It would appear that with the accelerating pace of automation, the time is approaching when many people will be among the permanently nonemployable and will become the recipients of government allowances. First industrialism robbed men of the pride in craftsmanship, and now it is robbing them of work itself. Without work, large numbers will be plagued by emptiness, ennui, a feeling of not belonging, a sense of guilt.

How will Americans use their ever-increasing leisure? In former times, they would have found outlets in family gatherings and in neighborhood, community, and church activities; but all these are now declining. Pursuits in the great outdoors, particularly hunting, fishing, and baseball on the green, were once the most common folk amusements in America; for obvious reasons these are disappearing. The American family has dwindled to the married couple and their minor children, and family life narrowly centers around financial security, the dwelling, the automobile, and the TV set. The multiplication of mechanical contrivances is divorcing

Americans from nature itself—and frequently from direct contacts with social reality. From time immemorial, a popular diversion has been to throng the market places, public squares, milling streets, and amusement areas in order to fraternize and take in the passing show of humanity. (This was once a practice in America, although it was always less extensive than in other countries.) Now, except for a flying visit by automobile to a supermarket, this is dying in America. Today, in many American cities after dark, a pedestrian is coming to be a curiosity, an object of police suspicion and barking dogs.

Foreign observers have pointed out repeatedly that because of the failure of secular humanism to penetrate the American population widely, there has been little folk appreciation in America of good art, music, drama, and literature. Even today, unlike Europeans, including Soviet citizens, few of the American folk find release in ballet, opera, concert, symphony, live theater, art museums, good books—and in the animated conversations which arise from such interests.

Again, foreign commentators have emphasized that the conception of the American as a breezy, gasconading hell-of-a-fellow is a frontier legend, a bucolic myth; that Americans are essentially an overworking, overworrying, overinhibited people; that they lack *gemütlichkeit*, the *joie de vivre*, a true gregariousness. Many Americans agree with this verdict. For instance, Nathaniel Hawthorne, after living among the English for four years and observing their society, at that time further into the industrial revolution than the American, concluded that Englishmen, from peer to peasant, "adhere closer to the original simplicity in which mankind was created than we ourselves do; they love, quarrel, laugh, cry, and turn their actual selves inside out with greater freedom than any class in America would think decorous."

Americans today have come a long way from their frontier, parochial past. Traditional puritan tabus, including many sexual ones, have weakened. Americans have become a more urbane and sophisticated people. They spend more money on recreations and amusements than any other people in the world. But they still lack gaiety, festive spirit, a sense of frolic. Their capacity for producing goods is still much greater than their capacity for enjoying life.

America's new sexual freedom is still self-conscious, often frenetic and flaunting, and not infrequently irresponsible. Our current literature of pot and pad, of navel and gonads, is unique in its voluminous outpouring, its lack of genuine ribaldry (where sex is merry and not dull), and its preoccupation with sexual techniques and the counting and classifying of orgasms. Even American drinking usually lacks conviviality; there is much moody solo drinking, and American bars are often downright sullen.

Now, there certainly is a great urge in present-day America to having "fun." In America, fun usually involves money, an automobile and other mechanical supports, and organization. However, to be "organized" does not mean to be truly gregarious. Foreigners observe that even in small social gatherings, Americans, particularly American women, seldom lose themselves, usually retain a certain narcissistic quality. And "fun" is not the same thing as "joy." Joy is pleasure heightened by spiritual élan; it largely depends on inner resources; it is spontaneous and unaffected; it may consist of the simplest activities; it may be individual or collective; and if collective it is festive and genuinely gregarious. European joy is for all ages, and all ages mix. American fun is segregated by age groups, and too much fun by the middle-aged and elderly is still considered indecorous.

Fundamentally, as Jules Henry points out, American fun represents a drive to "impulse release" from the tensions of advanced technicalization, a drive which frequently miscarries. For instance, our new ostentation in sex is in great measure a charade, a tease, "an abortive provocation," which leaves many Americans frustrated and foreigners puzzled.

A friend of mine, a lifelong resident of Helsinki, recently toured America. At the end of his visit he said to me: "What a wonderful country America is. What energy, what wealth, what comfort, what luxury, and to outward appearances what hedonism. It is all simply overwhelming! And yet there is something missing, terribly missing—a sense of joy."

VII

As society after society enters the machine age, the significant disparities between them are likely to be less and less in their formal institutions and more and more in their cultural capacity for making an affluent society a good society. Here, history may be preparing some surprises, for the country which pioneered the technocratic society may not be in the forefront in molding a spiritually satisfying one.

It is even possible that some of the societies we now call "underdeveloped" will in the next half century evolve more attractive ways of life than the advanced technocracies. In one way or another, the underdeveloped countries, through machinery and applied science, will modernize their economic life; but most of them are not likely to become fully technicalized societies. Thus they may overcome the hardships of the preindustrial society without submitting to the disciplines required in a thoroughly technicalized one, or losing much of the spontaneity, élan, and zest of people in simpler societies.[3]

Of those societies which have emerged or are emerging as genuine technocracies—the United States, Western Europe, the Soviet Union, China, and Japan—there is perhaps greater likelihood that Western Europe, because of its traditional *joie de vivre* and secular humanism, will escape many of the dehumanizing effects of the machine. There are Europeans who would insist that some countries in Eastern Europe—Hungary, Czechoslovakia, Poland, even Russia itself—have also been participants in the old humanism of Europe, and that in the end, despite Communism's narrow materialism, the basic cultural inheritance is likely to prove more influential in molding the way of life in these countries than the "more superficial" economic and political systems.

Currently there is an exciting intellectual renaissance in Poland, stimulated by the clash of Marxism, Catholicism, and existentialism. When professional and literary journals in Warsaw, Prague, and Budapest review scholarly American books, it is curious to note that not infrequently there is a questioning of the lengths to which American social scientists are carrying their quantitative methodologies. There has recently come out of Poland a book by Stanislaw Lem, *The Robot's Fables*, which is a satire on the cybernetic, computerized society and its reverence for arcane mathematical techniques. It is significant that this fantasy has emerged from a Communist country, and that it has not a grim Kafkaesque quality but instead a delightful Alice in Wonderland flavor.

China has a rich indigenous cultural heritage, and indeed China has played much the same role in fashioning Oriental civilization that Europe has played in fashioning the Occidental. In the end, after industrializa-

3. The goal of adopting machine technology enough to mitigate the hardships of a preindustrial society but not enough to rob the folk of the simple human enjoyments is one that Lewis Mumford would understand. Mumford in his *Technics and Civilization*, calls the period of the seventeenth and early eighteenth centuries in Western Europe "the eotechnic civilization." Despite the persistence of material hardships, the Commercial Revolution had brought many material improvements. Life was beginning to be comfortable. Values were still intensely human—not overmaterial, not overrational, not mechanical. Mechanical improvements were still prized because they saved labor and gave more time and energy for enjoyment. Throughout life, among rich and poor, the spirit of play was understood and fostered. Mumford writes: "The goal of the eotechnic civilization as a whole was not more power alone but a greater intensification of life."

tion, China's over-all cultural inheritance may prove more decisive than the Communist ideology in affecting the basic characteristics of Chinese life.

Although we are unaccustomed to thinking of long-term developments in this way, it is even possible that Japan and the United States, where cultures are more synthetic and further removed from their indigenous origins, will have the greatest difficulties transforming their technocracies into spiritually satisfying civilizations. The very qualities which made the Americans the pioneers of the technocratic society may operate to impede their converting that society into the good society.

It is unfortunate that America has no authentic conservative tradition and so few genuine conservatives. At this juncture in our history, an authentic conservatism would be alerting us to the Orwellian overtones in our society and to other invasions of privacy, would be skeptical of mere bigness and of technological innovation as the measure of progress, and would counsel a slowing down of population growth and some restraining of the pace of technological change. With science and technology dissolving the ancestral values and eroding religious certainties, a genuine conservatism, as Lippmann reminds us, would be in search of meaning which binds human experience and engages anew man's faculties and passion, of new forms in which enduring truths and values can be carried on in a world which is being rapidly transformed.

But what passes for American "conservatism" today is a literary and romantic Burkeanism; or a professional, fanatical anti-Communism; or Goldwaterism, which is a latter-day Horatio Algerism. The mainstays of Goldwaterism are the country-club sets, John O'Hara types along the cocktail circuits: materialists to the hilt, conspicuous and invidious consumers, brash go-getters who would, if they could, reduce even the rain and sunshine to profit-making gadgeteering.

Since the Americans pioneered the technicalized society (a historical achievement not without grandeur and even nobility), it is natural that the incipient evils of such a society should show themselves earlier in America than elsewhere. There they stand in bolder relief, too, for the American technocracy, unlike Communist technocracies, has no official materialistic philosophy and no formal antilibertarian institutions to obscure their technocratic origin. But it is well to remember that technocracy is new, that many evils now barely discernible or merely anticipated may fail to develop as technocracy matures, and that others may be conditioned and softened, and some eliminated, by conscious effort.

The American society is amazingly complex, has a myriad of facets, and contains many different groups. Indeed, here and there in that society are groups which in one way or another defy every generalization about the American culture suggested in this article. All this gives the American society flexibility, and it would be foolhardy to predict that the Americans will be unable to transform their technocracy into a spiritually satisfying civilization.

Americans in considerable numbers are aware of the challenge to privacy and liberty posed by the Orwellian tendencies in their new society. The threats to creativity implicit in the distorted trend to quantification have not yet materialized. In fact, in recent decades American creativity in the arts, literature, and theoretical science has been markedly increasing. But as every historian of civilizations realizes, past and borrowed cultural reservoirs may be diminishing at the very time a society outwardly appears most brilliant. However, atrophy may never set in if the learned societies, the foundations, and the federal government widen their programs for a revival and vast extension of the humanities and humanist values.

The possibilities for developing a rich mass culture surely exist. The motion picture has evolved into a sophisticated mass art, and television may eventually follow the same course. Hi-fi sets are making good music available to all classes in even remote nonmetropolitan areas. The trend to paperback books may yet work a cultural revolution.

During the past year, I have had occasion to visit many university campuses and to talk with hundreds of college students throughout the country. Members of the rising generation have an urgent sense of personal engagement in politics. They are in rebellion against the impersonality of the "multiversity." They are questioning the technocratic values of the whole society. They doubt that the quantitative methodologies to which they have been subjected are sufficient for exploring and understanding life. And frequently they say in effect: 'The Europeans, through machine technology and American business methods, are reducing their old evils of class distinction and widespread poverty. Now, if they can retain their old virtues—humanism, the sense of personality and privacy, reflection, relaxed ways of living—they will have a really superior civilization. With the Europeans taking on the best of American culture, why can't we Americans take on the best of European culture?" This cuts to the heart of the matter; but it reveals a too-facile view of the way cultural processes actually operate, and it minimizes the difficulty of alleviating traditional ills without at the same time impairing traditional virtues.[4]

VIII

Everywhere, under the impact of machine technology, and under a wide variety of regimes and "systems," old class and racial barriers are falling; and man, despite the population explosions, is developing the means to conquer the ageless material ills of preindustrial societies. Emphatically, the battle for material security is not yet won; but its winning is at last in sight. The vital concerns of the future will increasingly revolve around cultural satisfactions, psychological and spiritual freedoms. It would be tragic irony if the tools man fashioned to break the material shackles of old themselves became new shackles.

4. What does this observation do to the proposal in this article that American pragmatism should be infused with European humanism? Can something like this be "engineered" when it does not evolve naturally? Is this proposal naively nonhistorical? Perhaps so. But Western civilization's humanism was never absent in America, only relatively weak; it is not a new importation but a heritage which needs to be reinvigorated and enlarged. Would this undermine the American virtues derived from pragmatism? Hardly, for America's pragmatic values are so tenacious that they can stand considerable chastening in order that the goals and satisfactions of Americans may be broadened. Would humanism undermine democracy? No more than democracy undermines humanism. Humanism and democracy are not incompatible. Humanist values in Western Europe survived the liberal, democratic, and socialist movements; they are having more difficulty surviving technocracy. The trouble with such cults in America as the New Humanism is that they are antidemocratic, snobbish, precious, and "hot house."

~ Surrealism, Revolution, Reform—Or What?

In their drives to a more egalitarian society, the political-minded among today's youthful rebels, both the New Reformers and the New Leftists, are in the American tradition; but in rejecting democratic methods and embracing revolutionary ones with emphasis on mass violence to attain their objectives, both little and big, the New Leftists are outside the American tradition. In the more general youth movement to jettison a middle-class life-style, the young dissidents represent something much larger than politics. Both in speech and personal behavior, the dissenters are in revolt against the inhibitions of our contemporary society. The older generation should not be misled by what it often regards as the "way-out" behavior, the idiosyncratic antics, and the grotesque exhibitionism of the younger generation. Very often these are mannerisms by which the young dissidents "thumb their noses" at the growing conformist and other-directed tendencies of our technocratic society.

The following article does not deal with the struggle of the dissenting students to achieve reforms within the colleges and universities themselves. This takes two directions. One would widen student participation in university decision-making and administration. This is in line with the American democratic tradition and will succeed. Students will be given a larger "piece of the action." The other involves the curriculum and would halt the trend to making undergraduate courses mere adjuncts of the graduate schools, the trend to overemphasis of methodology, tools, and techniques. The students are asking for more courses which concern themselves with substance, existential experience, and comparative examination of values. In effect, the students are groping toward a more humanistic education. If this could be achieved, its repercussions throughout our society might be vitally beneficent in modifying the "soullessness" of technocracy; but victory in this aspect of the student revolt is much in doubt, for it runs against the grain of America's cultural tradition.

■ OUR CONTEMPORARY youth movements, both outside and inside the colleges, have aroused a passionate reaction in the United

States. There is much confusion in the public mind about the many elements that compose these movements, their temper, and their aims.

An important source of the confusion is the central duality that runs through them: they combine the criticisms of basic institutions, which characterize "major revolutions," with experiments in personal life-styles, which characterize "minor" ones.

Major revolutions, like the Reformation, the Cromwellian, the French, and the Russian, had a coherent body of revolutionary doctrine, and they drastically transformed basic religious, economic, social, and political institutions.

However, there have been "lesser" revolutions, those concerned chiefly with changes in family, sex, morals, and style. The nineteenth-century movement to emancipate women was one of these; feminist leaders resorted to all sorts of bizarre antics to draw attention to themselves; even the most sedate defended the right of Amelia Bloomer to wear her pantaloons on Boston Common.

Another such "side" revolution occurred in the United States in the 1920s in the otherwise placid years of Calvin Coolidge, when the parents and grandparents of today's student dissenters were themselves the young rebels who broke with America's puritan past; discovered Havelock Ellis and Freud; devoured Mencken, Nathan, Fitzgerald, and Percy Marks; exulted in jazz, the speak-easy, the petting party, and the parked roadster; and encouraged the belle to become the flirt, the flirt the flapper, and the flapper the "baby vamp."

The present-day youth movements are historically curious. They are calling for basic institutional changes, and at the same time they are spearheading drives to widen individual freedom in conduct, morals, manners, and life-styles and participating in a personal contraceptive revolution. Even in their criticisms of economic and political institutions, they have little or no fundamental philosophy, little or no body of doctrine.

To the uninitiated, these dissenting movements appear to be a jumble of the views of Thoreau, Sorel, De Sade, Guevara, Gandhi, Marcuse, Malcolm X, Bertrand Russell, Timothy Leary, Frantz Fanon, Martin Luther King, and Jean Genêt; a conglomeration of looking backward and looking forward, of violence and nonviolence, of idealist and nihilist, of integrationist and black nationalist, of transcendentalist and syndicalist, of seeker and hooligan, of hippy and leather-jacket, of aging beatnik and perennial youthmonger. On first impression, one is tempted to call this seething turbulence—wide-ranging, many-faceted, inchoate, kaleido-scopic, inarticulate, pragmatic, romantic, and contradictory—the Surreal-ist Revolution. Even the plazas, hangouts, bars, and bistros where the young rebels gather, in a bewildering array and disarray of dress and hairdos, representing the various youth roles and cults, have a surrealistic physical appearance.

II

More important off the campuses than on them, there are the hippies, the black-leather-jackets, and kindred groups. These were preceded, a decade ago, by the beatniks. Today's hippies are larger in number than were the beatniks, and they represent a less select group. They have as yet produced no literature comparable to that of the beatniks. Like the beatniks, they seek escape from the technocratic society and discovery of simpler and more personalized life-styles. Unlike them, however, they have a sense of mission and want to set an example for the rest of us, to point the way to a less complicated and involved existence. They are a softer lot than were the beats, and they would influence individuals and society by the tactics of gentleness and love. There are genuine uplifters among them who actually live the ideals they profess, but the movement has grown so large that the hippy ranks now contain many poseurs, ad-venturers, impostors, drifters, and moochers.

Certain "emancipated" areas in the big cities—like Haight-Ashbury in San Francisco, Oldtown in Chicago, the East Village in New York, and Jackson Square in New Orleans—have become hippy meccas. These big-city "free" neighborhoods, and others like them in smaller cities, have attracted college drop-outs and youngsters recently graduated from high schools, some of them sincerely imbued with the ideals of the move-ment, and some drawn to such communities for "kicks." There they live in small communal colonies where they share the expenses of "pad and chow." To meet these expenses of communal living, some work at odd jobs, some get money from home, some panhandle, and some plainly "free-load" on their brethren.

The hippies exalt love, yet most of them know little of love in the cre-ative, sacrificial, and redemptive sense; they claim that poverty and ascet-icism constitute the good and self-reliant life, yet they could not exist an instant except for the gratuities of an affluent society; they say they want to found communal rural colonies and dig their livings from the good earth, yet they cling tenaciously to the metropolitan environment; they assume the role of the gentle seeker, yet they often operate more like a band of begging gypsies.

The leather-jacket "hellions," another major group, pretend to be daring roisterers forever on the lookout for a game of "chicken," an affray, or a bold sex encounter; yet they ornament themselves with bracelets, anklets, pendants, and earrings, wear the regalia even when they have never owned a motorcycle, and one suspects they prefer group sex the better to stimulate an ambiguous sex drive.

Both groups resurrect stereotypes and mystiques from the pre-technocratic past—the hippies imitate the impoverished ascetic, the earnest seeker, the barefoot mudsill of the piney-wood South, the buckskin pioneer of the frontier West, the communal phalangist of the nineteenth century; the leather-jackets emulate the swashbuckler, the buccaneer, the eighteenth-century macaroni, the nineteenth-century Limehouse hooligan, the turn-of-the-century Paris apache—yet they all insist they have been uninfluenced by history and that history has nothing to teach them.

Most participants in youth cults, if they do not reject religion entirely, reject it in its conventional forms; they say instead they are seeking a more introspective knowledge of the self and a wider angle of vision of external reality—yet members of an "in" group will gather together, and, induced by physical and psychic excitations of one kind or another, including splashes of psychedelic colors on the walls, they will seek to "blow the brain" and turn the self "inside out" by swaying, moaning, and groaning to the rhythmic beat of a drum, reminiscent of the primitive religious camp meeting.

All the youth groups pride themselves on their nonconformity—yet the dress of each subgroup is slavishly imitative; there is a common catechetical jargon to condemn the middle class and "the establishment" and extol dissent; and all overuse the same stale five or six obscenities and four-letter words, which have lost all meaning and capacity to impress. This poverty of the lascivious imagination is all the more striking when compared to the wide-ranging, discriminating, spontaneous, colorful, and metaphorical obscenities of bucolic youth of the nineteenth century.

These young people fancy they are free spirits in matters of sex—yet ironically their sex attitudes express technocratic values; they have reduced sex to "experience" and "behavior"; they have robbed it of romance, playfulness, ribaldry, and joy; they duly enumerate and classify sex techniques; and their "in" authors reflect the methods of a Kinsey rather than the spirit of a Boccaccio, a Rabelais, or a "Tom Jones."

The crowning paradox is the way many of this younger generation, despite the obvious inconsistencies in their own behavior, have made an over-all fetish of "open honesty." The blessed words are sincerity, honesty, integrity; the damning words are dishonesty, cant, hypocrisy. It would appear, according to this rite of "absolute integrity," that good or evil resides not so much in the behavior itself as in the honesty or dishonesty with which one behaves; Jean Genêt and even De Sade, it is claimed, are "good" because they lack hypocrisy; hence almost any behavior would seem to be acceptable if it is practiced overtly, and care is taken not to deceive oneself or others.

III

There is much confusion about the relationship of hippies and dissenting college students. Most college-student dissenters are not outright hippies, but a number are. However, college students, even those who are not dissenters or activists, have been influenced by hippies, leather-jacket boys, older beatniks, and youthmongers. Even conservative college groups have been somewhat affected in their dress, manners, morals, and styles. Many a college student who appears quite conventional on campus will steal away on a Friday or Saturday night to an area where the "emancipated" gather, there to play the role of hippy or leather-jacket. Nor has there been an escape from some of the anomalies and contradictions which characterize the youth movements generally.

But when mature college dissenters embrace some of the attitudes of off-campus dissenters, there is a wider and deeper involvement with values, more serious attempts to be constructive, more concern about a quality civilization, more anxious searching for answers to the eternal problems of human existence. There is an even greater sense of mission, too; for these college youngsters believe they are spearheading drives which in the future will result in softening, liberalizing, and humanizing our materialistic, creature-comfort-oriented, overcompetitive, overorganized, monolithic-tending, other-directed technocracy.

Even the urge of some college students to take psychedelics appears to be somewhat different from that of the ordinary hippy or hooligan. With the latter, it seems largely sensual, but with the college student it seems to be a desire to widen perceptivity. Huston Smith has observed that most student pot-takers know that drugs solve nothing, but they believe

279

that psychedelics can magnify, mythify, at times clarify, and above all multiply the ways in which a problem or situation is viewed; and that this leads them to believe that they have a wider angle of vision than their elders and encourages them to relativize life-styles, perceiving the traditional as only one among a number of options the pot-takers feel they have virtually seen, not just imagined.

When accused of too wide an experimenting with sex, a member of the current college generation often replies something like this: "Yes, we probably experiment more widely in sex than did the younger generations of the past, but this is the inevitable result of the contraceptive revolution, and any younger generation first confronted with this revolution would have responded with a like widening of actual participation. This contraceptive revolution has enormous implications, some of which are not yet discernible, but one thing already seems clear—that women are emerging from womanhood to personhood, with the same latitude of personhood men have always enjoyed." Avant-garde students, then, more often view the changing sex mores in larger and truer perspective than do the off-campus cultists.

IV

All of the active dissident students on our American college and university campuses today constitute, when taken together, only a small minority of the total student enrollments. The dissenters fall roughly into three groups.

First, there are those with little interest in politics. These are primarily interested in divesting themselves of puritan, middle-class, and technocratic values; in encouraging individualistic, eccentric, and idiosyncratic behavior; in living life from day to day with spontaneity and zest; in making an oblique cultural attack on the technocratic society; in demonstrating that there are ways beyond the institutional to circumvent "1984"; in showing that the human spirit will "out" despite an intimidating environment.

Second, there are the political-minded who believe in working for reform within the system. These are the New Liberals or New Reformers, who agree that the activities of the first group may well have a liberating influence on our society but hold that an oblique cultural attack is not enough, that in addition we need specific political and institutional reforms to produce the Good Society.

Third, there are the New Leftists. These feel that the abuses in the

American society are beyond mere reformation, that a social revolution is necessary. Most observers agree that the New Leftists constitute a very small percentage of the college dissenters. However, the New Leftists are formidable beyond their numerical strength, for they are often guided by mature professional revolutionaries from the outside, and their sensational and explosive tactics capture the television cameras and the newspaper headlines.

There is a wide difference of opinion about the depth and permanent significance of these youth movements. Those who think of themselves as followers of Arnold Toynbee's interpretation of history see mounting evidence of cultural promiscuity, the disappearance of an authentic lifestyle, the taking on of the vulgar and barbarous morals, manners, and speech of the "internal proletariat"—sure signs of the disintegration of a civilization. Seymour Lipset plays down the youthful dissent when he argues that the majority of American college students is basically conservative. Robert E. Fitch goes beyond Lipset; he feels that this majority is even more conservative than the rest of the citizens of the country. On the other hand, Michael Novak and Huston Smith hold that change, and only change, is the basic issue in American higher education today; that it is a question of how far change should go. Do we believe that our capitalist democracy can secure freedom and justice for all, or must there be a serious rearrangement of the bases of power, wealth, and prestige? On which side do the universities stand, on the side of revolution or of reform within the system?

My own view is that what appears to be a cultural promiscuity is a brave attempt to reassert the human spirit in the face of an omnivorously encroaching technocracy; and that stripped of their fringes, the youth movements are much less surrealist than at first blush. I believe that Novak and Smith are correct in observing that the political dissenters roughly shake down into two opposing groups, the revolutionaries and the reformers; but I disagree that they have equal significance, for there is no revolutionary situation in the United States. I agree with Lipset and Fitch that the majority of American college students is conservative; but I also feel that it is the dissenters, through the New Reformers, who probably will make the most impact on the American politics of the future, set its pace and tone. Among the youthful reformers a vital creative minority is in the making, a minority that has something constructive to say to our people and is laying the groundwork for a new reform movement in America. The American society, which has a genius for periodically renewing itself, will likely give this reform movement considerable heed.

The general public tends to think of all student dissenters as New Leftists. It is high time that the aims and methods of the New Leftists, numerically a small group, and those of the New Reformers, a much larger one, are differentiated.

V

Unlike the Old Leftism in the 1930s, which was pro-Soviet, the New Leftism has lost faith in Russia as a revolutionary power. New Leftists feel that the Soviet Union, despite its revolutionary talk, has evolved from a revolutionary to a stabilizing force. The Soviets, in turn, view the New Leftists as pawns of Maoist China.

New Leftists are more highly organized than the liberal reformers. Students for a Democratic Society is a wide-ranging group of many faces, but essentially they seek social revolutionary ends by appeals to students, workers, intellectuals, and minority groups, particularly the Afro-Americans. The Afro-American militants themselves have a number of "black-power" organizations. These increasingly look upon Students for a Democratic Society as a white organization, and co-operation of black and white militants has become more and more difficult. Anarcho-syndicalism has an appeal to all the extremist groups, but increasingly the "black power" groups are emphasizing not anarcho-syndicalism so much as black nationalism, separate and autonomous black enclaves within American institutions and American communities. However, even organizations with seemingly common aims find co-operation difficult because of the rivalry of the leaders of the various groups. (On the other hand, liberal reformers and moderates, along with conservatives, are rallying against the destructive tactics of the extremists.)

The revival by the New Left of anarcho-syndicalism is sheer romance. Even around 1910, when Georges Sorel's anarcho-syndicalism was at the height of its influence, most revolutionaries chose the Red flag over the Black flag, and the vast majority of radicals rejected as unrealistic anarcho-syndicalism's basic thesis—that both the state and specialized management could be eliminated and control of industries simply turned over to the labor unions. Today, the technocratic revolution has made that thesis even more unrealistic. Back around 1910, too, Sorel's brilliant exaltation of absolute violence as a source of flaming revolutionary esprit and heroism was better poetry than politics, as is Frantz Fanon's similar exhortation for today.

An even more common sentiment than anarcho-syndicalism that runs across all of today's New Leftist groups, but entwining with anarcho-syndicalism's glorifying of violence, is the mystique of guerrilla warfare. The New Left's heroes are Mao the guerrilla theoretician, Ché Guevara the guerrilla "activist," and Frantz Fanon the passionate guerrilla rhetorician. Fanon has combined guerrilla warfare, anarcho-syndicalism, and Sorel's romanticizing of absolute violence into an inspiration to revolutionary élan, a guide to revolutionary action, and a vehicle of the revolt of the world's nonwhites against the whites.

True, Fanon makes a general appeal to the peoples of the underdeveloped countries, as victims of neocolonialism, but since most of these are nonwhites, his philippics have decidedly racial overtones. Fanon views the Third World as seething with revolution and uniting in action, but in reality that world is one of fierce rival nationalisms and tribalisms, one in which Asians are not above prejudices against Africans.

Another New Left hero is Herbert Marcuse, who insists that while the road to man's liberation is through science and machine technology, the decisive factor is what classes will control that technology. Socialism is still the answer, he says, and the workers should be the bearers of the revolution; but the trouble is that manual labor is declining, in the United States organized labor has become the handmaiden of the Establishment, and the new working producers, the technicians, are conservative and on the way to becoming an elite. Marcuse contradicts Fanon and maintains that the preconditions for the liberation and development of the Third World must emerge in the advanced industrial countries, that the chain of exploitation must be broken at its strongest link. The revolution is most likely to come as the result of breakdown in the United States, the result of growing contradictions within America's corporate capitalism. Meantime, let guerrillas nibble away at imperialism in the underdeveloped areas, and let our intellectuals, students, and minority groups prepare the values of the new society that will emerge, a society in which Materialism and Culturalism, Work and Art, will fuse.[1]

At bottom, Marcuse is a European intellectual with a nostalgia for

1. Marcuse's writings have had a beneficent influence in bringing to the attention of American youth the importance of humanist values. but his insistence that these can come to full flower for the first time in man's history only after there has been a drastic reconstruction of society is misleading. After the bitterness and turmoil of a revolution such as he envisages, humanist values in America would almost certainly be weaker rather than stronger.

the pre-World War I simplistic socialism shot through with old humanist values and the *joie de vivre*. He finds encouragement in the May 1968 student rebellion in France, in which the earlier naive socialism and anarcho-syndicalism were infused with the old European humanism. (At Toulouse, striking students even demanded the revival of the language of the Troubadours.) Frail evidence, indeed, in the face of the monumental fact that collectivism in the Soviet Union has resulted in an authoritarian bureaucracy, a technocratic elite, a new puritanism, and Socialist Realism in the arts. (So one turns to Maoist China, where the revolution is still inchoate enough to allow one to dream of an ideal order.)

Now, American New Leftist students have little familiarity with theory, and little grounding in Europe's cultural tradition of humanism, but they have a burning sense of injustice and are attracted to the tactics of violence and guerrilla warfare. This attraction is mostly moonshine.

Why is most of this moonshine? For one thing, the student New Leftists do not really intend to employ absolute violence and conduct guerrilla warfare; their seizing college buildings and indulging in miscellaneous sabotage are about the actual extent of it. For another thing, though they meant it, they should know that even in our revolutionary age, guerrillas alone have never made a revolution, and they provide no disciplined doctrine or organization with which to create a revolutionary state. The Bolsheviks won in Russia only after several years of civil war in which highly organized Red armies fought highly organized White armies. Guerrilla activities were merely ancillary. The Mexican revolutionaries of around 1910, Mao in China, and Castro in Cuba, all in different ways, were successful because they were able to utilize forces and conditions beyond mere guerrilla warfare. For another thing, it would appear that today the trend in the world is not to revolution; the new nations in the underdeveloped areas, no matter what the particular "ism" formally avowed, are increasingly turning from ideology to machine technology as panacea; this is one of the major reasons why more and more of the new nations have resorted to army elements to govern them, for the army in an underdeveloped nation contains most of the country's technicians. For still another thing, violence begets violence, and unless there is a genuine revolutionary situation, as there is not in most of the underdeveloped countries today, and certainly not in the United States, resorts to widespread violence lead only to needless blood-baths and suppressions. In Vietnam, where there has long been a genuine revolutionary situation, guerrilla warfare has made sense, but even there the main thrusts, first against the French and then against the Americans, have come from the well-organized regular armies of Ho Chi Minh.

VI

The New Reformers, unlike the New Leftists, aim not at revolutionary changes but at reform within the system. However, the New Reformers claim that today the conventional methods of getting political results are not enough. They feel that the tenacity of white racism requires shock treatment. They say that a quarter of a century of Cold War has created hard and unrealistic popular stereotypes that the politicians of all parties are afraid to dispel, and that these stereotypes play into the hands of the military-industrial complex, which finds them profitable. They also contend that the control of the mass media is in the hands of "the establishment," of which the military-industrial complex is an important part, and that the voices of dissent do not get an adequate or a fair hearing, unless sensational methods are used.

Hence the New Reformers, too, have resurrected some of the more direct methods of the past to make their views effective. One of these is a revival of transcendentalism, the moral assertion of the individual conscience against the established policies of organized society. Another is the resort to massive nonviolent civil disobedience. The heroes of the New Reformers are Martin Luther King and Mohandas Gandhi—particularly Gandhi, because of his unshakable belief that "love force" or "soul force," expressed by a transcendental faith in the higher law of our being, the voice of the individual conscience, and applied through nonviolent civil disobedience, can make a revolution. Now, reason the liberal reformers, if these are sufficient to win a revolution, they certainly must be sufficient to effectuate reforms.

In this looking back to the methods of an earlier age, the New Reformers, like the New Leftists, have not escaped romanticism. For transcendentalism is a rarer quality today than it was earlier, because in the past there were more individuals who were inner-directed, frequently impelled by religious conviction; but today religious fervor has declined, and for a number of other reasons even the most sturdy individualists find it difficult to resist the pressures of their other-directed environment. More important, in historical perspective, nonviolent civil disobedience has more often been mystique than successful reality, for when applied on a large scale it has almost always led to violence, as Gandhi discovered at Amritsar and during the various civil disobedience campaigns he waged

285

during the 1930s. But Gandhi was spared the necessity of ever having to acknowledge this publicly, for he *won* his revolution—largely because the old imperialism had played out, and the British were wise enough to see this and withdrew of their own volition before they had used a fraction of their force.

The evolution of the transcendentalist views of Henry David Thoreau are most pertinent. When at last Thoreau became convinced that slavery would never be abolished by peaceful and nonviolent agitations, he supported violence. Unlike Gandhi, conditions forced Thoreau to acknowledge the limitations of nonviolence to achieve a transcendent moral goal. Of all the Abolitionists, Thoreau was the most extreme (far more than Garrison, whom he derided) in his defense of John Brown's bloody violence at Pottawatomie, and at Harper's Ferry, which Thoreau called "a sublime spectacle." When Thoreau saw that "love force" or "soul force" would not be enough to win the revolutionary objective of freeing the slaves, he argued in effect that "brute force" is transmuted into "soul force" when moral man is driven to violence in pursuit of a transcendent moral cause dictated by his inner conscience.

Most people think of Thoreau's essay on "Civil Disobedience" as his mature view of nonviolence; they have failed to follow the development of his thought on this subject during the subsequent years, which climaxed in his passionate "A Plea for Captain John Brown." Contrary to the stereotype, it is not a common confidence in nonviolence that links Gandhi and Thoreau; rather it is their common faith in the "higher law" of transcendentalism.

New Leftists sometimes co-operate in civil disobedience campaigns with liberal reformers and other moderates, not only to achieve an immediate common objective, but also because professional revolutionaries, who insinuate themselves among young New Leftists, know that nonviolent civil disobedience may in fact come out of hate as well as love, and that massive civil disobedience may often escalate into violence and create situations favorable to extremists by embarrassing liberals.

Considering the delicate nature of nonviolent civil disobedience, the achievements of such men as Martin Luther King, Benjamin Spock, and William Coffin, backed by legions of students, have been remarkable. These men, and those like them, stand in the transcendentalist tradition of the primitive Christians, the Lollards, the Hussites, the Anabaptists, the Quakers, and the Abolitionists. And they employed nonviolent civil disobedience on a large scale more successfully and with less actual violence than has been usual in history. We must not blink the fact, however, that these contemporary transcendentalists succeeded largely because "the establishment," correctly viewing the situation as actually nonrevolutionary, refrained from taking massive countermeasures which, had they been taken, would probably have led to widespread violence and bloodshed.

Recognizing all this, the more aware of the New Reformers feel an especial urgency about formulating and effectuating a timely program of reform to prevent a revolutionary situation from developing in the United States. They know that reformers, in the middle, are always in peril of being smothered by revolutionaries on the one hand and reactionaries on the other; and that today's nonviolent transcendentalists included in their ranks want desperately to escape the dilemma of the nonviolent transcendentalists of Thoreau's day—the dilemma of having to support bloody violence or to lapse into a passivity that might well weight the scales against achieving their own goals of future betterment.

VII

The lineaments of a New Reform movement in American politics are taking shape in the minds of dissident liberal students. It is already clear that the attitudes and proposals of the New Reformers will differ in a number of ways from those of the liberal wing of the Democratic party of the past quarter of a century.

These youthful dissenters will carry with them a pronounced suspicion of the military. They will dedicate themselves to dispelling the myth of an international Communist monolith and to limiting American military commitments abroad. They will press to reduce the swollen defense and space budgets, so that large sums may be diverted from the military and from outer-space projects to foreign economic aid, and especially to large increases in spending at home in order that we may ameliorate critical domestic racial and social problems. They will claim that this can be done without serious risks to the national security. They will be less concerned than were the older reformers with placating business groups, and more concerned with bypassing intermediate layers of middlemen to assure that the public spending is spread widely, really gets to the people it is designed to help, actually provides the services intended. They will be as zealous to put checks on bureaucrats, brain-trusters, social workers, and professional planners as they will be to restrain antisocial behavior in private business and the professions. Above all, they will insist that bread-and-butter measures, standard-of-living goals, are not enough;

that the psychic needs and spiritual satisfactions of a people must be taken into much larger account.

Up to this time, the New Reformers have not decisively and vividly differentiated themselves from the New Leftists. New Reformers sometimes co-operate with New Leftists to win a common immediate objective, and on occasion liberals join leftists on some collateral issue while continuing to differ from them on the main issue. For instance, when outside police or national guardsmen are called in to quell New Leftist disorders, New Reformers often stand with New Leftists to vindicate the autonomy of the campus and to badger the "intruders," no matter what their differences on the main issue which initially triggered the crisis. However, when New Reformers co-operate with New Leftists, they should take care to make plain their reasons for doing so, always preserving their own identity; for if the New Reformers are ever to have the impact on the American politics of the future they now think they will have, they must be vigilant to emerge clearly in the public mind as a separate, responsible, and persuasive entity in their own right.[2]

---

2. A number of adult liberals have not dissociated themselves sufficiently from the New Leftists. Certain of these, among them some well-known journalists, supported the militants in their fights to establish at various universities virtually independent black-studies departments dedicated not to scholarship but to revolutionary agitation. Even more damaging, in the trial of the Chicago Seven, some of these same journalists slanted their news stories in such a way as to make heroes of the New Left defendants, despite their bizarre disturbances to disrupt the court proceedings, and to make the judge, who carried on despite these disturbances, a villain. It was certainly consistent with liberalism to question the statute under which the defendants were tried and the wisdom of indicting them; but once their trial got under way, a show of sympathy for disruptive tactics in the courtroom was in effect to sanction attempts to sabotage the judicial process and the rule of law itself. Here was an opportunity for liberal reformers to rebuke the wayward liberal journalists, but too few did so. Unless the liberal reformers make a better record of differentiating themselves from the revolutionaries, their bright political prospects for the future will wither.

❧ Index

Abdul Karim Kassem, 58

Academicians in U.S.: changing character, 222, 233, 234; advanced degrees no guarantee of quality, 222; feuding, 224–225; evaluating teaching, 225; teaching survey courses, 225–226; detachment from community, college, and students, 226–227; teaching stimulus to teacher's creativity, 227; contacts with students outside classroom, 227; preoccupation with research, 227–228; television teaching, 228; "professional" humanists, 228; the professional associations, 229; publishing, 229–231; rackets, 231–232; foundations and grants, 231–233; wangling a grant, 232–233; rewards academic life, 233. *See also* Colleges and universities

Acheson, Dean, 161

Acton, Lord John, 186

Adams, Brooks, 261

Adams, Henry, 201, 261

Addams, Jane, 247

Adenauer, Konrad, 97, 161

Adler, Mortimer, 243

Aiken, Senator George D., 15

Alcorn, Governor James L., 151–152

Alien and Sedition laws, 100

Alliance for Progress, 164

Altgeld, Peter, 247

America First, 214

American foreign policy: stereotyped American opinion of world politics, 38–50; threat to unity Western alliance, 42, 43, 50; failure differentiate foreign "isms," 44, 248; misunderstanding economic and social nature anti-imperialist revolutions, 44–45; overemphasis armaments and neglect economic aid programs, 45; long-time blindness national rifts within Communism, 45–46; ignoring basic historical trends, 46–47; failure distinguish Hitlerian and Communist threats, 47; legalistic and moralistic approach, 48; pressures internal national minorities, 48; neglect consult allies, 49; contrast former British world leadership, 49–50; misconceptions mixed economies in underdeveloped countries, 63–64, 76; big power brinkmanship but not war, 66, 79; measures taken supernuclear powers to maintain balance of power and freeze status quo, 66–68; interventions to maintain status quo, 66–67, 259n; nonintervention to change status quo, 67, 75, 79, 259n; why thaw Cold War, 73–80; exaggeration explosive forces perpetuates crisis mentality, 80–83; dangers right-wing pressures, 81–82; failure rely on Soviet-Chinese rivalry as stabilizer of Asian balance of power, 81n; bipolarization gives way to depolarization, 83; Kennedy foreign policy, 160–166. *See also* Kennedy, John F.; Nuclear revolution

American presidency: President popular tribune, 104; agitation for restraints in conduct foreign policy, 104; liaison with Congress and direct communication with public, 104; treachery opinion polls, 104–105; evaluating advice of "experts," 105; screened information by aides, 105; controlling bureaucracy, 105, 112–113; various presidential roles, 105–106; overshadowed by Congress until twentieth century, 106; some strong Presidents prior twentieth century, 106–107; growth under each twentieth-century President, 107–109; reasons for, 110–112; still popular tribune despite democratization national House, 111 and n; powers inadequate meet responsibilities, 112–113; powers not self-executing, 113; proposed abolition electoral college, 133; proposed nominating primary, 133–134; proposed six-year term, 133–134. *See also* Kennedy, John F.; Presidential nominating conventions; Wilson, Woodrow

Anarcho-syndicalism, 282

Angell, Sir Norman, 68

289

Index

Index

Deminform, 31

Democracy. *See* Centralization; Functional groups; Intellectuals; Liberalism

De Sade, the Marquis, 279

Dewey, John, 229, 233, 238, 247

Dewey, Thomas E., 138, 139

Dickens, Charles, 175, 191, 196, 198, 201, 210

Dix, Dorothea, 247

Dos Passos, John, 115

Douglas, Stephen A., 140, 141

Dulles, John Foster, 37, 164

Dumas, Alexander the elder, 222

Dunavan, Robert, 218

Easton, David, 242

Einstein, Albert, 6, 267

Eisenhower, Dwight D., 49, 88, 109, 135, 139, 142, 154, 155

Eismann, Bernard, 162n

Elections in U.S.: Agitation for presidential nominating primary, 133–134, 145 and n; for direct election President, 133–134; prospect multiplying number presidential campaigns, 133–134; growing intensity and costs electioneering, 133–134, 146–148; trend to presidential elections becoming national landslides, 144; state and local primaries, 144–145

Eliot, T. S., 240, 243

Ellis, Havelock, 276

Emerson, Ralph Waldo, 237, 247

Erskine, John, 228

European Common Market, 161, 162

Fanon, Frantz, 52, 282, 283

Farley, James A., 138

Faulkner, William, 167

Federalism in U.S. *See* Centralization; Nationalism

Fine, Benjamin, 91n

Fitch, Robert E., 281

Fitzgerald, F. Scott, 276

Flappers, 276

Fleming, D. F., 125

Folk hero in history, 168–169

Fourteen Points, 119–120

Fox, Charles James, 6, 8, 31

Franco, General Francisco, 258

Franklin, Benjamin, 247

Frémont, John C., 140, 141

Freud, Sigmund, 116, 219, 267, 276

Functional groups in politics: part played in developing "bourgeois" national states in Europe, 10, 25; possible operation within an international Communist state, 11, 12; probable operation within an effective United Nations, 13; failure to operate internationally and why, 17, 28–30, 32–34; probable eventual emergence within national Communist states, 20; development of an international state dependent on group alliances cutting across nations, 23–25; group alliances as makers of American federalism and nationalism, 25–28; United Nations structure discourages international group politics, 28–30; ideological affinities but not genuine group politics cut across nations, 30–32; impact on twentieth-century centralization in U.S., 93, 99, 101–102; part played in democratizing presidential nominating conventions, 143–144, 145; in nationalizing American politics, 144; autonomous private associations as libertarian buffers, 186–187; basis countervailing power, 249

Galbraith, John K., 96, 249–250

Gandhi, M. K., 59, 63, 285–286

Garfield, James A., 141, 144

Garner, John Nance, 122, 138

Garrison, Lindley M., 116

Genêt, Jean, 279

George, Alexander, 116

George, Henry, 247

George, Juliet, 116

Glass, Carter, 124

Goldwater, Barry, 132, 149, 185, 211

Goldwaterism, 170, 171–173, 181, 184–185, 272

Gompers, Samuel, 124

Goodwin, Richard N., 157

Gothic architecture, 207

Grant, U. S., 140

Grants-in-aid to states, 90, 92

Greeley, Horace, 141, 247

Guerrilla warfare, 284–285

Guevara, Ché, 283

Hagerty, James C., 139

Hale, William Bayard, 115

Hamilton, Alexander, 247

Hancock, Winfield S., 140, 141

Harding, Warren G., 108, 137, 141

Hargis, Billy James, 180

Harrison, Benjamin, 141, 144

Harrison, William Henry, 140, 141

Hartz, Louis, 173, 182, 245, 246 and n, 247, 248

Harvey, George, 123

Index

Index

McAdoo, William G., 124
MacArthur, Douglas, 49
McCarthy, Eugene, 132
McCarthy, Joseph, 152
McCarthyism: too personalized conception of, 37; made possible by American alarm over world situation, 37; European reaction to and misconceptions of, 38–39; why Western Europeans escaped similar movement, 39–44, 49; strained unity of Western alliance and confidence in American leadership, 42, 43, 50; Kennedy's evasion of issue, 152–153; mentioned, 100, 109, 171, 211
McCombs, William F., 123
McCumber, Porter B., 119
McKinley, William, 128, 247
Machine technology: characteristics English and American societies in early industrial age, 191–192; today's mass society product of advanced mechanization, and its ambivalence, 250–252 and n; trend to technocracy dominant today, 255; meaning of term, 255; how differs from earlier industrialism, 255; technocratic values, managerial society, and administrative state cutting across all ideologies, 256–258; technocracy not anticipated by Socialists, 256; Communism in practice becomes managed state capitalism, 256–257; underdeveloped countries in various ways developing machine technology, 257–258; trend away from ideology to machine technology as panacea, 257–258; even strongly traditionalist countries moving to mechanization, 258; conquest outer space, 258; nuclear revolution a phase of larger technological revolution, 259–260; pressures to revolutions and wars may yield to technological advances, 260; effective machine technology in underdeveloped countries dependent on combination of factors, 260–261; early warnings technocracy's threats, 261; difficulties controlling science and machines to socially desirable ends, 261–262; Orwellian threats, 262–263; new technocratic elite and prospect of caste society, 263–264; trend to quantification, 264–265; inbreeding technocratic values and elites threat to continued vitality science and technology, 267; increase in both tensions and leisure, 268; problems in U.S. of leisure time, 268–269; further decline in U.S. of folk gregar-

iousness, 269; weakness in U.S. of aesthetic pleasures secular humanism, 269; decline in U.S. of puritan tabus but continuing absence of *joie de vivre*, 269–270; American drive to "fun" distinguished from joy, 270; reconciling technocratic and nontechnocratic values differs from country to country, 270–274; aspiration in underdeveloped countries to combine machine technology with traditional folkways, 270–271 and n; hardships preindustrial societies giving way to new concerns, 274. *See also* Humanism; Nuclear revolution
Maddox, Lester, 185
Madison, James, 24, 36
Mallan, James P., 152n
Manion, Clarence, 180
Mann, Horace, 247
Mansfield, Mike, 133
Mao Tse-tung, 21, 283. *See also* China
Marcuse, Herbert, 52, 283–284 and n
Marks, Percy, 276
Marshall Plan, 71–72
Martin, Ralph G., 153n
Martineau, Harriet, 127, 201
Marx, Karl, 179, 267
Marxism-Leninism: traditional view nationalism, 7–8, 17; concessions to nationalism, 10–11, 18, 19; traditional view internationalism, 7–8, 9, 10, 17; miscarriage Marxist internationalism, 19; views on colonialism and anticolonialism, 53; how older Marxism differed from younger Leninism on anticolonial revolutions, 60–61, 77; impact on colonial peoples of miscarriage Marxism in Western Europe, 60–61, 76–77; failure most anti-imperialist revolutions to merge with Communist revolutions, 73; revisionism whittling away original doctrines, 78–79; technological interpretation history causes acceptance military restraints to nuclear revolution, 81; in practice has resulted in nationally managed state capitalisms, 256–257. *See also* Communism; Imperialism and anti-imperialism; Internationalism; Nationalism; Nuclear revolution
Mason, George, 172
Mboya, Tom, 59
Melville, Herman, 207
Mencken, H. L., 115, 181, 243, 276
Menninger, Karl A., 115
Mill, John Stuart, 186

Index

Wilson, Woodrow (*Cont.*)
century ideology Western democracies, 120; his breakthrough to internationalism, 120–121 and *n*; oratory, 121–122; give-and-take with politicians, 122; relations with Congress, 122; efficient administrator, 123; breaks with political lieutenants, 123–124; Paris Peace Conference, 124–125; Lodge Reservations, 125; exaggeration personal traits, 125–127, 126*n*; reasons for hypercritical approach to him, 127–130, 127*n*; subjected to psychoanalysis, 130–131; target of the "new politics," 131
Wisotsky, Isidore, 218
Wobblies, 213, 215
Wolfe, Thomas, 238
World Federalists, 23, 29, 36

World Federation Trade Unions, 31, 34

Youth movements: mentioned, 211; question technocratic values, 273–274; campus reforms advocated by, 275; New Leftists, New Reformers, and New Stylists differentiated, 275, 280–288; flaming youth of 1920s, 276; today's incoherence in, 276–277; beatniks, hippies, and leather-jackets, 277–278; contradictions, 277–279; campus and noncampus offbeat groups, 279–280; psychedelics, 279–280; contraceptive revolution, 280; significance youth movements, 281–282; activities and mystiques New Leftists, 282–285; mystiques and program New Reformers, 285–288 and *n*

DATE DUE

Library